Mathematical Sciences and Applications

PROCEEDINGS OF INTERNATIONAL CONFERENCE ON MATHEMATICAL SCIENCE AND APPLICATIONS (ICMSA 2023)

Edited by

Prof. Sanjay Chaudhary
Department of Mathematics, Dr. Bhimrao Ambedkar University Agra, India, 282002

Prof. Sanjeev Kumar
Department of Mathematics, Dr. Bhimrao Ambedkar University Agra, India, 282002

Dr. Shyamli Gupta

Department of Mathematics, Dr. Bhimrao Ambedkar University Agra, India, 282002

CRC Press
Taylor & Francis Group
Boca Raton London New York

CRC Press is an imprint of the
Taylor & Francis Group, an **informa** business

First edition published 2024
by CRC Press
4 Park Square, Milton Park, Abingdon, Oxon, OX14 4RN

and by CRC Press
2385 NW Executive Center Drive, Suite 320, Boca Raton FL 33431

CRC Press is an imprint of Informa UK Limited

British Library Cataloguing-in-Publication Data
A catalogue record for this book is available from the British Library

ISBN: 9781032588612 (pbk)
ISBN: 9781003451808 (ebk)

DOI: 10.1201/9781003451808

Typeset in Sabon LT Std
by HBK Digital
Printed and bound in India

DR. BHIMRAO AMBEDKAR UNIVERSITY, AGRA

Prof. Ashu Rani
Vice Chancellor

(B⁻ NAAC Accredited)

Paliwal Park, Agra

Department of Mathematics Dr. Bhimrao Ambedkar University, Agra has been recognized as a Centre of Excellence by the State Government. With financial assistance from the State Govt. and the University, the Department of Mathematics organized a 3-day International Conference on Mathematical Sciences and Applications during March, 24-26, 2023.

Now the proceeding of the conference with some selected and referred presentations is being published by Taylor & Francis Publications.

As Vice-Chancellor of the university, I feel honoured to present these proceedings to the mathematics community in and outside India. I am sure these research papers would encourage and ignite young minds to explore mathematical ideas and techniques a few steps ahead.

(Prof. Ashu Rani)

Dedication

Prof. J. B. Shukla
(1937 – 2023)

Prof. Shukla was an Indian mathematician who specialised in modelling of ecological, environmental, physiological, and engineering systems.

He was awarded in 1982 the Shanti Swarup Bhatnagar Prize for Science and Technology. Other prestigious awards include the FICCI Award, and the Distinguished Service Award in Mathematical Sciences from Vijnana Parishad of India in 1997. Prof. Shukla did significant work on biofluid dynamics, in particular peristaltic transport of faeces in intestines and on the interaction of biorheological aspects of blood flow and arterial stenosis. He also made contributions in the areas of population dynamics and the mathematical theory of epidemics.

Prof. Satya Deo
(1944 – 2022)

Prof. Satya Deo collapsed during an award function just after receiving the Life Time Achievement award in the field of Mathematics from Vijñāna Parishad of India. He had been Vice Chancellor of APS University, Rewa and R. D. University, Jabalpur; and also served the NASI in the capacity of the General Secretary, Treasurer and Chief Editor of National Academy Science Letters.

Prof. Deo was the elected Fellow of the NASI in 1991 and the President of the Indian Mathematical Society in 1999. He was a recipient of the Distinguished Service Award by the Mathematical Association of India and the Indian Science Congress Gold Medal by the Prime Minister of India.

Prof. P. N. Pandey
(1952 – 2022)

Prof. Pandey worked as Vice Chancellor, NGB Vishwvidyalaya, Prayagraj. He also worked as a Professor & Head, Department of Mathematics, University of Allahabad, Allahabad. He was the Founder Fellow & General Secretary, of the International Academy of Physical Sciences. His field of interest were Differential Geometry, Bio-fluid-Mechanics, Vedic Mathematics, Mathematical Modelling and Bioinformatics. He guided around 30 D.Phil. and 1 post doc. He received several awards. IAPS was his dream to serve the society in different fields of physical sciences.

Contents

List of Figures

List of Tables

Preface

The papers appearing in these proceedings are part of talks, oral presentations and poster presentations given at the *International Conference on Mathematical Sciences and Applications* held in the Department of Mathematics, Dr. Bhimrao Ambedkar University, Agra (India) from March 24-26, 2023. The Conference was held under the auspices of the Mathematics Department which is recognized and founded by the U.P. State Govt. as a Centre of Excellence in Mathematics. The aim of the conference was to have a gathering of experts from the different field of Mathematical sciences and its applications in physical and biological sciences. The conference attracted more than 250 participants from almost every State of the Union of India and a few from abroad. Out of around 150 presentations, only selected and refereed papers have been included in these proceedings.

The papers included represent several aspects of proven and applied mathematics, and interdisciplinary mathematics, Statistics, Computer Science, Physics, Chemistry, Biology and Medical and environmental science.

The authors have taken great pain in writing, rewriting and modifying these papers. The referees have spent considerable time in judging and rejudging the quality of papers. It is hoped that these contributions would enrich readers knowledge and motivate many to take up challenging problems which may contribute to the growth of the subject and welfare of the society.

We thank all those who have been associated, directly or indirectly with the publication of the Proceedings.

Prof. Sundar Lal
Ex Vice Chancellor
VBS Purvanchal University, Jaunpur
Chairman ICMSA-2023

List of Reviewers

S.No.	Name of Reviewer	Affiliation
1	Prof. Sundar Lal	Dr. Bhimrao Ambedkar University, Agra
2	Prof. S.N. Singh	T.D.P.G. College, Jaunpur
3	Prof. Narendra Kumar	ICFAI University, Jaipur
4	Prof.Rakesh M Patel	Gujrat Arts and Science College, Ahmdabad
5	Prof.Anand Joshi	Lucknow University, Lucknow
6	Prof.Rashmi Bhardwaj	Indraprastha University, Delhi
7	Prof. Deepak Kumar	Manav Rachna International University, Faridabad
8	Prof. N.K. Sharma	IIMT Group of Institutions, Noida
9	Prof. Rajesh Saini	Bundelkhand University, Jhansi
10	Prof. Atul Chaturvedi	P.S.I.T., Kanpur
11	Prof. S.P. Singh	D.E.I. Dayalbag , Agra
12	Prof. Avanish Kumar	Bundelkhand University, Jhansi
13	Prof. Sanjit Kumar	LNCT, Bhopal
14	Prof. Alpana Mishra	Sharda University, Noida
15	Prof. Radha Gupta	Dayanand Sagar Univ., Banglore
16	Dr. Neha Mishra	NJILM, ICMR, Agra
17	Prof. Kirtiwant Ghadle	Dr. Babasaheb Ambedkar Marathwada University, Aurangabad
18	Prof. Sanjay Jain	St. John's College, Agra
19	Prof. Ratnesh Ranjan Saxena	DDU College, Karampur, Delhi
20	Prof. Bhupendra Rana	CCS University, Merrut
21	Dr. Akhil Mittal	Govt. Science College, Santrampur, Gujrat
22	Prof. Mamta Nigam	DDU College, Karampur, Delhi

1 A note on q-series and Clausen type identities

Shyamli Gupta[a] and Manish Gokani

Department of Mathematics, Institute of Basic Science, Khandari Campus, Dr. Bhimrao Ambedkar University, Agra-282004, Uttar Pradesh, India

Abstract

Bailey transformation leads to a significant impact on the growth of transformation theory of hypergeometric series and number theory. In the present work, a widely used transformation established by Bailey, is used to established some interesting Clausen type identities.

Keywords: q-hypergeometric series, q-hypergeometric series with multiple bases and Clausen type identities

1. Introduction

In 1828, Clausen has discovered that a specific hypergeometric function $_2F_1$'s square can be represented as a hypergeometric function $_3F_2$. Later, Ramanujan (1914) exploited special examples of Clausen's identities in his derivation of the 17 series for $1/\pi$. Since then numerous initiatives to uncover fresh parallels of Clausen's identities have been made. Utilizing the research on the modular forms and modular functions as inspiration, three new Clausen's identities were established by Chan et al. (2021). Also, Denis, Singh and Singh (2022) have established certain Clausen type identities with the help of a known series identity.

In 1936 Bailey, gave the following result

$$\sum_{m=0}^{\infty} \beta_m \, \delta_m = \sum_{m=0}^{\infty} \alpha_m \gamma_m \tag{1}$$

Where, $\beta_m = \sum_{r=0}^{m} \alpha_r u_{m-r} v_{m+r}$

and $\gamma_m = \sum_{r=m}^{\infty} \delta_r u_{r-m} v_{r+m}$.

The aforementioned transformation produces a number of results that are significant in the fields of hypergeometric series transformation theory and number theory. By utilizing Bailey's transformation, some intriguing Clausen type identities has been established in the current paper.

2. Definitions and notations

Definitions and notations used in the paper are as follows:

According to the definition, a generalized basic hypergeometric series with two bases,

---

[a]shyamlig@gmail.com

$$\underset{A+B}{\square}\phi_{C+D}\begin{bmatrix}(a);(b);q,q_1;z\\(c);(d);q^i,q_1{}^j\end{bmatrix}=\sum_{n=0}^{\infty}\frac{[(a);q]_n[(b);q_1]_n z^n q^{i\binom{n}{2}}q_1{}^{j\binom{n}{2}}}{[q;q]_n[(c);q]_n[(d);q_1]_n} \tag{2}$$

where the series of A parameters $a_1, a_2, \dots a_A$ is denoted by (a).

Also, for $|q| < 1$ and arbitrary a

$[a; q]_n \equiv (1; a)(1 - aq)(1 - aq^2) \dots (1(aq^{n-1}), n>0$

and $[a: q]_0 = 1$

Further, $\binom{n}{2} \equiv n(n - 1)/2$.

A convergence occurs in the right-hand side of series (2) for $|q|, |q_1|$ less than one and $|z|$ less than infinity, when i and j both are greater than 0 and $|q|,|q_1|, |z|$ less than one when i, j both are 0. In that case q^i and $q_1{}^j$ are dropped from the notation.

Also,

$$\underset{A}{\square}\phi_B\begin{bmatrix}(a);q;z\\(b)\end{bmatrix}=\sum_{n=0}^{\infty}\frac{[(a);q]_n z^n}{[q;q]_n[(b);q]_n} \tag{3}$$

valid for $|z| < 1, |q| < 1$.

The following known summations of shortened q-series were also employed in the present work.

$$\underset{2}{\square}\varphi_1\begin{bmatrix}\alpha, x; q; q\\\alpha x q\end{bmatrix}_N=\frac{[\alpha q, xq;q]_N}{[q, \alpha xq;q]_N} \tag{4}$$

(cf. Agarwal R.P, (1978))

$$\underset{4}{\square}\phi_3\begin{bmatrix}\beta, q\sqrt{\beta}, -q\sqrt{\beta}, y; q; \frac{1}{y}\\\sqrt{\beta}, -\sqrt{\beta}, \frac{\beta q}{y}\end{bmatrix}_K=\frac{[\beta q, yq;q]_K}{[q, \beta q/y;q]_K y^K} \tag{5}$$

(cf. Agarwal R. P, (1978))

$$\underset{6}{\square}\phi_5\begin{bmatrix}\beta, q\sqrt{\beta}, -q\sqrt{\beta}, y, z, t; q; q\\\sqrt{\beta}, -\sqrt{\beta}, \beta q/y, \frac{\beta q}{z}, \frac{\beta q}{t}\end{bmatrix}_N=\frac{[\beta q, yq, zq, tq;q]_N}{[q, \frac{\beta q}{y}, \frac{\beta q}{z}, \frac{\beta q}{t};q]_N} \tag{6}$$

with $\beta = yzt$

(cf. Agarwal R. P, (1978))

$$\underset{3}{\square}\phi_2\begin{bmatrix}a, b, q; q; q\\e, f\end{bmatrix}_N=\frac{(e-abq)(q-e)}{(aq-e)(e-bq)}\left\{1-\frac{[a,b;q]_{N+1}}{[e/q, abq/e;q]_{N+1}}\right\} \tag{7}$$

provided $ef = abq^2$

(cf. Agarwal R. P, (1978))

$$\sum_{r=0}^{m} \frac{(1=\beta p^r q^r)[\beta;p]_r[y;q]_r y^{-r}}{(1-\beta)[q;q]_r[\beta p/y;p]_r} = \frac{[\beta p;p]_m[yq;q]_m y^{-m}}{[q;q]_m[\beta p/y;p]_m}$$

(8)

(cf. Gasper and Rahman (1991))

3. Main outcomes

The results developed in this research are as follows:

$$\Box_2\phi_1\begin{bmatrix}\alpha, x; q_1; q_1\\ \alpha x q_1\end{bmatrix}\Box_2\phi_1\begin{bmatrix}a, y; q; q\\ ayq\end{bmatrix}$$

$$={}_4\phi_3\begin{bmatrix}a, y; \alpha q_1, x q_1; q, q_1; q\\ ayq; q_1, \alpha x q_1\end{bmatrix}+\frac{(1-\alpha)(1-x)q_1}{(1-q_1)(1-\alpha x q_1)}{}_4\phi_3\begin{bmatrix}aq, yq; \alpha q_1, x q_1; q, q_1; q_1\\ ayq; q_1^2, \alpha x q_1^2\end{bmatrix}$$

$$-{}_4\phi_3\begin{bmatrix}a, y; \alpha x; q, q_1; q q_1\\ ayq; q_1, \alpha x q_1\end{bmatrix}$$

(9)

$$\Box_2\phi_1\begin{bmatrix}\alpha, x; q_1; q_1\\ \alpha x q_1\end{bmatrix}\Box_4\phi_3\begin{bmatrix}\beta, q\sqrt{\beta}, -q\sqrt{\beta}, y; q; \frac{1}{y}\\ \sqrt{\beta}, -\sqrt{\beta}, \frac{\beta q}{y}\end{bmatrix}$$

$$=$$

$$\Box_6\phi_5\begin{bmatrix}\beta, q\sqrt{\beta}, -q\sqrt{\beta}, y; \alpha q_1, x q_1; q, q_1; \frac{1}{y}\\ \sqrt{\beta}, -\sqrt{\beta}, \frac{\beta q}{y}; q_1, \alpha x q_1\end{bmatrix}+\frac{(1-\alpha)(1-x)q_1}{(1-q_1)(1-\alpha x q_1)}\times$$

(10)

$$\Box_4\phi_3\begin{bmatrix}\beta q, yq; \alpha q_1, x q_1; q, q_1; \frac{q_1}{y}\\ \frac{\beta q}{y}; q_1^2, \alpha x q_1^2\end{bmatrix}\qquad -{}_6\phi_5\begin{bmatrix}\beta, q\sqrt{\beta}, -q\sqrt{\beta}, y; \alpha, x; q, q_1; \frac{q_1}{y}\\ \sqrt{\beta}, -\sqrt{\beta}, \frac{\beta q}{y}; q_1, \alpha x q_1\end{bmatrix}$$

$$\Box_2\phi_1\begin{bmatrix}\alpha, x; q_1; q_1\\ \alpha x q_1\end{bmatrix}\Box_6\phi_5\begin{bmatrix}\beta, q\sqrt{\beta}, -q\sqrt{\beta}, y, z, t; q; q\\ \sqrt{\beta}, -\sqrt{\beta}, \frac{\beta q}{y}, \frac{\beta q}{z}, \frac{\beta q}{t}\end{bmatrix}=$$

$$\Box_8\phi_7\begin{bmatrix}\beta, q\sqrt{\beta}, -q\sqrt{\beta}, y, z, t; \alpha q_1, x q_1; q, q_1; q\\ \sqrt{\beta}, -\sqrt{\beta}, \frac{\beta q}{y}, \frac{\beta q}{z}, \frac{\beta q}{t}; q_1, \alpha x q_1\end{bmatrix}+$$

(11)

$$\frac{(1-\alpha)(1-x)q_1}{(1-q_1)(1)\alpha x q_1)}\times_6\phi_5\begin{bmatrix}\beta q, yq, zq, tq; \alpha q_1, x q_1; q, q_1; q_1\\ \frac{\beta q}{y}, \frac{\beta q}{z}, \frac{\beta q}{t}; q_1^2, \alpha x q_1^2\end{bmatrix}$$

$$-{}_8\phi_7\begin{bmatrix}\beta, q\sqrt{\beta}, -q\sqrt{\beta}, y, z, t; \alpha, x; q, q_1; q q_1\\ \sqrt{\beta}, -\sqrt{\beta}, \frac{\beta q}{y}, \frac{\beta q}{z}, \frac{\beta q}{t}; q_1, \alpha x q_1\end{bmatrix}$$

with $\beta = yzt$

$$
{}_2\Box\phi_1\begin{bmatrix}\alpha, x; q_1; q_1 \\ \alpha x q_1\end{bmatrix}\ {}_3\Box\phi_2\begin{bmatrix}\beta, y, q; q; q \\ z, t\end{bmatrix} =
$$

$$
{}_5\Box\phi_4\begin{bmatrix}\beta, y, q; \alpha q_1, x q_1; q, q_1; q \\ z, t; q_1, \alpha x q_1\end{bmatrix} + \frac{q_1(1-\alpha)(1-x)(e-\beta y q)(q-z)}{(1-q_1)(1)\alpha x q_1)(\beta q - z)(z - y q)}
$$

$$
\times \left\{ {}_3\Box\phi_2\begin{bmatrix}\alpha q_1, x q_1, q_1; q_1; q_1 \\ q_1^2, \alpha x q_1\end{bmatrix} - \frac{(1-\beta)(1-y)}{(1-z/q)(1-\beta y q/z)}\ {}_5\Box\phi_4\begin{bmatrix}\beta q, y q, q; \alpha q_1, x q_1; q, q_1; q_1 \\ z, \frac{\beta y q^2}{z}; q_1^2, \alpha x q_1^2\end{bmatrix} \right\}
$$ (12)

$$
- {}_5\Box\phi_4\begin{bmatrix}\beta, y, q; \alpha, x; q, q_1; q q_1 \\ z, t; q_1, \alpha x q_1\end{bmatrix}
$$

provided $zt = \beta y q^2$

$$
{}_2\Box\phi_1\begin{bmatrix}\alpha, x; q_1; q_1 \\ \alpha x q_1\end{bmatrix}\ {}_3\Box\phi_2\begin{bmatrix}y; \beta q p; p; q, q p, p; \frac{1}{y} \\ -; \beta; \frac{\beta p}{y}\end{bmatrix} = {}_5\Box\phi_4\begin{bmatrix}y; \beta q p; \beta; \alpha q_1, x q_1; q, q p, p, q_1; \frac{1}{y} \\ -; \beta; \frac{\beta p}{y}; q_1, \alpha x q_1\end{bmatrix}
$$

$$
+ \frac{(1-\alpha)(1-x)q_1}{(1-q_1)(1)\alpha x q_1)}\ {}_4\Box\phi_3\begin{bmatrix}y q; \beta p; \alpha q_1, x q_1; q, p, q_1; \frac{q_1}{y} \\ -; \frac{\beta p}{y}; q_1^2, \alpha x q_1^2\end{bmatrix} - {}_5\Box\phi_4\begin{bmatrix}y; \beta p q; \beta; \alpha, x; q, q p, p, q_1; \frac{q_1}{y} \\ -; \beta; \frac{\beta p}{y}; q_1, \alpha x q_1\end{bmatrix}
$$ (13)

4. Proof

Taking $u_r = 1 = v_r$ in the identity (1), after some simplification we obtain

$$
\sum_{m=0}^{\infty} \alpha_m \sum_{r=0}^{\infty} \delta_r = \sum_{m=0}^{\infty}\left(\sum_{r=0}^{n} \alpha_r\right)\delta_m
$$
$$
+ \sum_{m=0}^{\infty} \alpha_{m+1}\left(\sum_{r=0}^{m} \delta_r\right) - \sum_{m=0}^{\infty} \alpha_m \delta_m
$$ (14)

Taking,

$$
\alpha_n = \frac{[\alpha, x; q_1]_n q_1^n}{[q_1, \alpha x q_1; q_1]_n},
$$

in (14) and using the summation (4), we obtain

$$
\sum_{n=0}^{\infty} \frac{[\alpha, x; q_1]_n q_1^n}{[q_1, \alpha x q_1; q_1]_n} \sum_{r=0}^{\infty} \delta_r
$$

$$
= \sum_{n=0}^{\infty} \frac{[\alpha q_1, x q_1; q_1]_n}{[q_1, \alpha x q_1; q_1]_n} \delta_n + \sum_{n=0}^{\infty} \frac{[\alpha, x; q_1]_{n+1} q_1^{n+1}}{[q_1, \alpha x q_1; q_1]_{n+1}}\left(\sum_{r=0}^{n} \delta_r\right)
$$ (15)

$$
= \sum_{n=0}^{\infty} \frac{[\alpha q_1, x q_1; q_1]_n}{[q_1, \alpha x q_1; q_1]_n} \delta_n + \frac{(1-\alpha)(1-x)q_1}{(1-q_1)(1)\alpha x q_1)}\sum_{n=0}^{\infty} \frac{[\alpha q_1, x q_1; q_1]_n q_1^n}{[q_1^2, \alpha x q_1^2; q_1]_n}\left(\sum_{r=0}^{n} \delta_r\right).
$$

Now, taking

$$\delta_n = \frac{[\beta, y; q]_n q^n}{[q, ayq; q]_n},$$

in (15) and using (4), we get our main result (9).

Again, setting

$$\delta_n = \frac{[\beta, q\sqrt{\beta}, -q\sqrt{\beta}, y; q]_n y^{-n}}{\left[q, \sqrt{\beta}, -\sqrt{\beta}, \frac{\beta q}{y}; q\right]_n},$$

in (15) and using (5), we get our main result (10).

Taking

$$\delta_n = \frac{[\beta, q\sqrt{\beta}, -q\sqrt{\beta}, y, z, t; q]_n q^n}{\left[q, \sqrt{\beta}, -\sqrt{\beta}, \frac{\beta q}{y}, \frac{\beta q}{z}, \frac{\beta q}{t}; q\right]_n}$$

with $\beta = yzt$ in (15) and using (6), we get our main result (11).

Next, taking

$$\delta_n = \frac{[\beta, y; q]_n q^n}{[z, t; q]_n}$$

with $zt = \beta yq^2$ in (15) and using (7), we get our main result (12).

Taking

$$\delta_n = \frac{(1 - \beta p^n q^n)[\beta; p]_n [y; q]_n y^{-n}}{(1 - \beta)[q; q]_n \left[\frac{\beta p}{y}; p\right]_n}$$

in (15) and using (8), we get our main result (13).

5. Conclusion

In this paper, some Clausen type identities are obtained by using suitable summation formulae in Bailey transformation. Several other results are also derived by using the similar approach.

References

1. Agarwal, R. P. (1978). Generalized hypergeometric series and its application to the theory of combinatorial analysis and partitions, (Unpublished monograph).
2. Almkvist, G., Straten, D. V. and Zudilin, W. (2011). Generalizations of Clausen's formula and algebraic transformations of Calabi–Yau differential equations. *Proc. Edinb. Math. Soc.*, 54, 273–295.

3.   Bailey, W. N. (1936). Series of hypergeometric type which are infinite in both directions. *Quart. J. Math. (Oxford)*, 7, 105–115.

4.   Brychkov, Yu. A. (2011). On power series of products of special functions. *Appl. Math. Lett.*, 24, 1374–1378.

5.   Chan, H. C. et al. (2011). New analogues of Clausen's identities arising from the theory of modular forms. *Adv. Math.*, 228, 1294–1314.

6.   Clausen, T. (1828). Under die Falle wenna die Reihe $y = 1 + \frac{\alpha.\beta}{1.\gamma}x + \cdots$ ein qudrat von der Form $x = 1 + \frac{\alpha'\beta'\gamma'}{1.\delta'e'}x + \cdots$ hat. *J. fur Math.*, 3, 89–95.

7.   Denis, R. Y. and Singh, S. N. (2002). Transformation of poly-basic hypergeometric functions. *Proc. Third Ann. Conf. S.S.F.A.*, March 4–6, 191–220.

8.   Denis, R. Y., Singh, S. N., and Singh, S. P. (2002). On reducibility of certain q-double hypergeometric series and Clausen type identities. *South East Asian J. Math. Math. Sci.*, 1(1), 1–17.

9.   Gasper, G. and Rahman, M. (1991). Basic hypergeometric series. Cambridge: Cambridge University Press.

10. Gupta, S. (2022). On Certain Continued fractions involving basic bilateral hypergeometric function $_2\Psi_2$. *South East Asian J. Math. Math. Sci.*, 18(2), 71–76.

11. Ramanujan, S. (1914). Modular equations and approximations to π. *Quart. J. Math. (Oxford)*, 45, 350–372.

2 Availability analysis of warm standby system with general repair times

Kanta[a] and Sanjay Chaudhary[b]

Department of Mathematics, Dr. Bhimrao Ambedkar University, Agra-282002, Uttar Pradesh, India

Abstract

In the present scenario, we determine the availability for a system that uses warm standby components. It is taken into account with switching failure and general repair times. This one exists presumptively that the amount of failure and repair times of main (primary) components as well as backup (standby) components are exponentially and usually distributed. When failing units are replaced with standby units, a significant possibility q of switch failure is predicted. The reboot delay occurs during the process of switching from a standby component to a primary component. With parameter β, the boot up (reboot) time is expected as exponentially. Using the supplemental variable technique and a Laplace transformation, availability is derived. We determine explicit formulas for the availability facilitating three distribution (repair time distributions), namely the exponential, gamma, and uniform, letting used for numerical comparison.

Keywords: Availability, general repair time, reboot delay, switching fault, Laplace transformation

1. Introduction

In present day scenario, availability is the most useful tool to know about any system, and how long it works. To improve the system's availability various repair facilities, standbys, and maintenance are used. Cold, warm, and hot standbys are the three types. The failure rate of the cold standby is zero. Warm standby components fail at lesser rate than primary components, but the hot standby components fail at the same rate as the primary components. Standby components are used to boost the system's availability. Cox (1955) proposed an effective supplemental strategy for converting a non-Markovian procedure to a Markovian process by using supplementary parameters. Wang and Chen (2009) examined the obtainability for different systems using varying reparation timeframes, boot up delays, and exchanging faults. Yen et al. (2020) evaluated system availability using detecting latency and regular repair periods of time.

2. System description

We analyze a serviceable system composed of one primary and three standbys through λ (failure rate of primary components) and α (failure rate of standby components) individually. This system has boot up (reboot delay) and failure of switching. This one exists presumptively that the amount of failure and repair

[a]singhkanta80@gmail.com, [b]scmibs@hotmail.com

times of main (primary) components as well as backup (standby) components are exponentially and usually distributed. In this system, standby units work as warm standby units, whenever, the primary unit is failed. Failures may occur during the switch from the standby component to the primary component. It is projected that there would be a considerable probability q of switching failure. We assume in the system that a reboot delay occurs during the process of switching from a standby component to a primary component. The reboot timings are expected to be distributed exponentially according to the variable β (Reboot delay rate). The primary as well as standby components are thought to be repairable as well. The repair times of the system's units are random variables (i.i.d.) with B (u) (repair distribution ($u \geq 0$)), b (u) (function of probability density ($u \geq 0$)), and b_1 (mean time repair). U denotes the amount of time required for the broken unit under repair, such as a supplemental variable. There are some other notations as following, $P_{x,y}(t)$ denotes probability at time t, where x and y are operating and warm standby units respectively, where x = 0, 1 and y = 0, 1, 2, 3 and $P_{x,y}$ (s)– Laplace-Stieltjes transformation of $P_{x,y}(t)$, $P_{x,y}^{*(1)}(s)$– a first-order derivative of $P_{x,y}(t)$ concerning s (Figure 2.1).

3. System equations

Steady-state equations for the system are:

$$0 = -(\lambda + 3\alpha)P_{1,3} + P_{1,2}(0) \tag{1}$$

$$0 = -\beta P_{0,3} + \lambda(1-q)P_{1,3} \tag{2}$$

$$-\frac{\partial}{\partial u}P_{1,2}(u) = -(\lambda + 2\alpha)P_{1,2}(u) + 3\alpha P_{1,3}(u) + \beta P_{0,3}(u) + b(u)P_{1,1}(0) \tag{3}$$

$$0 = -\beta P_{0,2} + \lambda q(1-q)P_{1,3} + \lambda(1-q)P_{1,2} \tag{4}$$

$$-\frac{\partial}{\partial u}P_{1,1}(u) = -(\lambda + \alpha)P_{1,1}(u) + 2\alpha P_{1,2}(u) + \beta P_{0,2}(u) + b(u)P_{1,0}(0) \tag{5}$$

$$0 = -\beta P_{0,1} + \lambda(1-q)P_{1,1} + \lambda q(1-q)P_{1,2} + \lambda q^2(1-q)P_{1,3} \tag{6}$$

$$-\frac{\partial}{\partial u}P_{1,0}(u) = -\lambda P_{1,0}(u) + \alpha P_{1,1}(u) + \beta P_{0,1}(u) + b(u)P_{0,0}(0) \tag{7}$$

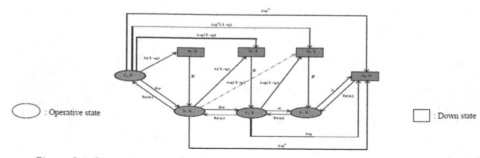

◯ : Operative state   ▢ : Down state

Figure 2.1 State transition diagram

$$-\frac{\partial}{\partial u}P_{0,0}(u) = \lambda q^3 P_{1,3}(u) + \lambda P_{1,0}(u) + \lambda q P_{1,1}(u) + \lambda q^2 P_{1,2}(u) \tag{8}$$

We define boundary conditions,

$$P_{1,3}(u) = b(u)P_{1,3} \tag{9}$$

$$P_{0,3}(u) = b(u)P_{0,3} \tag{10}$$

$$P_{0,2}(u) = b(u)P_{0,2} \tag{11}$$

$$P_{0,1}(u) = b(u)P_{0,1} \tag{12}$$

Following are the results from equations (1), (2), and (4).

$$P_{1,2}(0) = (\lambda + 3\alpha)P_{1,3} \tag{13}$$

$$P_{0,3} = \frac{\lambda(1-q)}{\beta}P_{1,3} \tag{14}$$

$$P_{0,2} = \frac{\lambda q(1-q)}{\beta}P_{1,3} + \frac{\lambda(1-q)}{\beta}P_{1,2} \tag{15}$$

Now taking Laplace-Stieltjes transformation of equation (3), and using equations (9) & (10).

$$(\lambda + 2\alpha - S)P_{1,2}^{*}(s) = 3\alpha B^{*}(s)P_{1,3} + \beta B^{*}(s)P_{0,3} + B^{*}(s)P_{1,1}(0) - P_{1,2}(0) \tag{16}$$

Setting S=λ+2α in equation (16) and substituting equations (13) & (14). We get,

$$P_{1,1}(0) = \phi P_{1,3}, \qquad \text{Where } \phi = \frac{(\lambda + 3\alpha) - B^{*}(\lambda + 2\alpha)[3\alpha + \lambda(1-q)]}{B^{*}(\lambda + 2\alpha)} \tag{17}$$

Setting S=0 in equation (16) and after putting equations (13), (14) & (17). We estimate the following result.

$$P_{1,2} = P_{1,2}^{*}(0) = \frac{(\lambda + 3\alpha)[1 - B^{*}(\lambda + 2\alpha)]}{(\lambda + 2\alpha)B^{*}(\lambda + 2\alpha)}P_{1,3} \tag{18}$$

On substituting equation (18) into equation (15), we obtain

$$P_{0,2} = \theta P_{1,3}, \qquad \text{Where } \theta = \frac{\lambda(1-q)}{\beta}\left[q + \frac{(\lambda + 3\alpha)\{1 - B^{*}(\lambda + 2\alpha)\}}{(\lambda + 2\alpha)B^{*}(\lambda + 2\alpha)}\right] \tag{19}$$

On taking Laplace-Stieltjes transformation of equation (5), and using equation (11). We obtain,

$$(\lambda+\alpha-S)P_{1,1}^{*}(s)=\beta B^{*}(s)P_{0,2}+2\alpha P_{1,2}^{*}(s)-P_{1,1}(0)+B^{*}(s)P_{1,0}(0) \tag{20}$$

Now put S=λ+α in equation (16) and substitute equations (13), (14) & (17). We obtain,

$$P_{1,2}^{*}(\lambda+\alpha)=\frac{(\lambda+3\alpha)\{B^{*}(\lambda+\alpha)-B^{*}(\lambda+2\alpha)\}}{\alpha B^{*}(\lambda+2\alpha)}P_{1,3} \tag{21}$$

Setting S=λ+α in equation (20) and putting the value of equations (17), (19) & (21). We get,

$$P_{1,0}(0)=\Pi P_{1,3} \tag{22}$$

Where $\Pi=\dfrac{1}{B^{*}(\lambda+\alpha)}\left[\phi-\beta\theta B^{*}(\lambda+\alpha)-2(\lambda+3\alpha)\left\{\dfrac{B^{*}(\lambda+\alpha)}{B^{*}(\lambda+2\alpha)}-1\right\}\right]$

Setting S=0 in equation (20) and substituting equations (17), (18), (19) & (22). We obtain,

$$P_{1,1}=\psi P_{1,3} \tag{23}$$

Where $\psi=\dfrac{1}{(\lambda+\alpha)}\left[\dfrac{\phi\{1-B^{*}(\lambda+\alpha)\}}{B^{*}(\lambda+\alpha)}+\dfrac{2(\lambda+3\alpha)}{B^{*}(\lambda+2\alpha)}\left\{\dfrac{\alpha\{1-B^{*}(\lambda+2\alpha)\}}{(\lambda+2\alpha)}+\dfrac{B^{*}(\lambda+2\alpha)}{B^{*}(\lambda+\alpha)}-1\right\}\right]$

This implies from after putting equations (18) & (23) into equation (6).

$$P_{0,1}=\frac{\lambda(1-q)}{\beta}\left[q^{2}+\psi+\frac{q(\lambda+3\alpha)\{1-B^{*}(\lambda+2\alpha)\}}{(\lambda+2\alpha)B^{*}(\lambda+2\alpha)}\right]P_{1,3} \tag{24}$$

On taking Laplace-Stieltjes transformation on both sides of equation (7), and exhausting equation (12). We have,

$$(\lambda-s)P_{1,0}^{*}(s)=\beta B^{*}(s)P_{0,1}+\alpha P_{1,1}^{*}(s)-P_{1,0}(0)+B^{*}(s)P_{0,0}(0) \tag{25}$$

Now setting S=λ in equation (16) and substitute equations (13), (14) & (17). We obtain,

$$P_{1,2}^{*}(\lambda)=\frac{(\lambda+3\alpha)\{B^{*}(\lambda)-B^{*}(\lambda+2\alpha)\}}{2\alpha B^{*}(\lambda+2\alpha)}P_{1,3} \tag{26}$$

We get the following result after setting S=λ in equation (20) and putting the values of equations (17), (19), (22) & (26). We get,

$$P_{1,1}^{*}(\lambda)=\frac{1}{\alpha}\left[\phi\left\{\frac{B^{*}(\lambda)-B^{*}(\lambda+\alpha)}{B^{*}(\lambda+\alpha)}\right\}+(\lambda+3\alpha)\left\{\frac{2B^{*}(\lambda)}{B^{*}(\lambda+\alpha)}-\frac{B^{*}(\lambda)}{B^{*}(\lambda+2\alpha)}-1\right\}\right]P_{1,3} \tag{27}$$

Setting S=λ in equation (25) and substituting equations (22), (24) & (27). We estimate

$$P_{0,0}(0) = \Omega P_{1,3} \tag{28}$$

$P_{0,0}(0) = \Omega P_{1,3}$

Where " $\Omega = \dfrac{1}{B^*(\lambda)}\left[\dfrac{\phi\{B^*(\lambda)-B^*(\lambda+\alpha)+1\}}{B^*(\lambda+\alpha)} - \beta\theta - \lambda(1-q)B^*(\lambda)\left\{q^2 + \psi + \dfrac{q(\lambda+3\alpha)\{1-B^*(\lambda+2\alpha)\}}{(\lambda+2\alpha)B^*(\lambda+2\alpha)} - \dfrac{2(\lambda+3\alpha)}{B^*(\lambda+\alpha)}\left(\dfrac{B^*(\lambda+\alpha)}{B^*(\lambda+2\alpha)}-1\right) + (\lambda+3\alpha)\left(\dfrac{2B^*(\lambda)}{B^*(\lambda+\alpha)} - \dfrac{B^*(\lambda)}{B^*(\lambda+2\alpha)} - 1\right)\right\}\right]$ "

Setting S=0 in equation (25) and substituting equations (22), (23), (24) & (28). We obtain

$$P_{1,0} = P_{1,0}^*(0) = \frac{1}{\lambda}\left[\lambda(1-q)q\left\{q + \frac{(\lambda+3\alpha)\{1-B^*(\lambda+2\alpha)\}}{(\lambda+2\alpha)B^*(\lambda+2\alpha)}\right\} + \{\lambda(1-q)+\alpha\}\psi + \Omega - \Pi\right]P_{1,3} \tag{29}$$

On taking Laplace-Stieltjes transformation on both sides of equation (8), and exhausting equation (9). We get,

$$sP_{0,0}^*(s) = P_{0,0}(0) - \lambda q^3 B^*(s)P_{1,3} - \lambda P_{1,0}^*(s) - \lambda q P_{1,1}^*(s) - \lambda q^2 P_{1,2}^*(s) \tag{30}$$

When we differentiate equation (30) with regard to S and set S=0, we obtain

$$P_{0,0}^*(0) = b_1\lambda q^3 P_{1,3} - \lambda P_{1,0}^{*(1)}(0) - \lambda q P_{1,1}^{*(1)}(0) - \lambda q^2 P_{1,2}^{*(1)}(0) \tag{31}$$

Where $b_1 = -B^{*(1)}(0)$,

After differentiate equation (16) with regard to S, at S=0, and substituting equations (14), (17), and (18). We get

$$P_{1,2}^{*(1)}(0) = \frac{1}{(\lambda+2\alpha)}\left[\frac{(\lambda+3\alpha)\{1-B^*(\lambda+2\alpha)\}}{(\lambda+2\alpha)B^*(\lambda+2\alpha)} - b_1\{\lambda(1-q)+3\alpha+\phi\}\right]P_{1,3} \tag{32}$$

Differentiate equation (20) with regard to S, at S=0, and substituting equations (19), (22), (23) & (32).

$$P_{1,1}^{*(1)}(0) = \frac{1}{(\lambda+\alpha)}\left[\psi - \frac{b_1}{(\lambda+2\alpha)}\{(\lambda+2\alpha)(\beta\theta+\Pi) + 2\alpha[\lambda(1-q)+3\alpha+\phi]\} + \frac{2\alpha(\lambda+3\alpha)\{1-B^*(\lambda+2\alpha)\}}{(\lambda+2\alpha)^2 B^*(\lambda+2\alpha)}\right]P_{1,3} \tag{33}$$

Similarly, we obtain by differentiate equation (25) with regard to S, setting S=0, and substituting equations (24), (28), (29), & (33).

$$P_{1,0}^{*(1)}(0) = \frac{1}{\lambda}\left[\begin{array}{l}\{q^2(1-q)(1-\lambda b_1) + \{(1-q)(1-\lambda b_1) + \frac{\alpha}{\lambda} + \frac{\alpha}{(\lambda+\alpha)}\}\psi + \frac{(\lambda+3\alpha)\{1-B^*(\lambda+2\alpha)\}}{(\lambda+2\alpha)B^*(\lambda+2\alpha)}\{q(1-q)(1-\lambda b_1) + \frac{2\alpha^2}{(\lambda+\alpha)(\lambda+2\alpha)}\} + \\ \frac{1}{\lambda}\{\Omega(1-\lambda b_1) - \Pi\} - \frac{\alpha b_1}{(\lambda+\alpha)(\lambda+2\alpha)}\{(\lambda+2\alpha)(\beta\theta+\Pi) + 2\alpha[\lambda(1-q)+3\alpha+\phi]\}\end{array}\right]P_{1,3} \tag{34}$$

Now we get the following result after substituting the equations (32), (33), & (34) into equation (31).

$$P_{0,0} = \xi P_{1,3} \tag{35}$$

Where

$$\xi = \left[\begin{array}{c}\left[q^2\{(1-q)(1-\lambda b_1)+q\lambda b_1\}-\{(1-q)(1-\lambda b_1)+\dfrac{\alpha(2\lambda+\alpha)+\lambda^2 q}{\lambda(\lambda+\alpha)}\}\psi-\dfrac{1}{\lambda}\{\Omega(1-\lambda b_1)-\Pi\}-\dfrac{(\lambda+3\alpha)\{1-B^*(\lambda+2\alpha)\}}{(\lambda+2\alpha)B^*(\lambda+2\alpha)}\{q(1-q)(1-\lambda b_1)+\dfrac{2\alpha(\alpha+\lambda q)+\lambda q^2(\lambda+\alpha)}{(\lambda+\alpha)(\lambda+2\alpha)}\}\right]\right] \\ +\dfrac{b_1(\alpha+\lambda q)}{(\lambda+\alpha)(\lambda+2\alpha)}\{(\lambda+2\alpha)(\beta\theta+\Pi)+2\alpha[\lambda(1-q)+3\alpha+\phi]\}+\dfrac{\lambda q^2 b_1}{(\lambda+2\alpha)}\{\lambda(1-q)+3\alpha+\phi\}\end{array}\right]$$

Using the following normalizing condition to obtain $P_{1,3}$ which is too ample to be shown here.

$$P_{1,3}+P_{1,2}+P_{1,1}+P_{1,0}+P_{0,0}+P_{0,3}+P_{0,2}+P_{0,1}=1 \tag{36}$$

It is assumed that three reboot delay states $(0, 3)$, $(0, 2)$, and $(0, 1)$ are system down states. Thus, we get the availability A_v as follows.

$$A_v = P_{1,3}+P_{1,2}+P_{1,1}+P_{1,0} \tag{37}$$

Substituting equations (18), (23) & (29) into equation (37), we obtain

$$A_v = \left[1+\dfrac{(\lambda+3\alpha)\{1+q(1-q)\}\{1-B^*(\lambda+2\alpha)\}}{(\lambda+2\alpha)B^*(\lambda+2\alpha)}+\left(2-q+\dfrac{\alpha}{\lambda}\right)\psi+\dfrac{1}{\lambda}\{\lambda(1-q)q^2+\Omega-\Pi\}\right]P_{1,3} \tag{38}$$

4. Special cases for A_v

We used three distributions for repair time in this work namely exponential, gamma, and uniform. The expressions of availability are given below.

I.   **Exponential distribution:** we set $b_1 = 1/\mu$ (mean repair time), where μ denotes the repair rate. We ensure the following values by taking Laplace transformation, $B^*(\lambda)=\dfrac{\mu}{\mu+\lambda}$, $B^*(\lambda+\alpha)=\dfrac{\mu}{\mu+\lambda+\alpha}$, $B^*(\lambda+2\alpha)=\dfrac{\mu}{\mu+\lambda+2\alpha}$.
   By equation (38), where Π_1, Ψ_1 & Ω_1 have the usual meaning as before for the exponential repair time distribution.

$$A_{VM} = \left[1+\dfrac{(\lambda+3\alpha)\{1+q(1-q)\}}{\mu}+\left(2-q+\dfrac{\alpha}{\lambda}\right)\psi_1+\dfrac{1}{\lambda}\{\lambda(1-q)q^2+\Omega_1-\Pi_1\}\right]P_{1,3} \tag{39}$$

II. **Gamma distribution:** With parameter r, we fix $b_1 = 1/\mu$ (mean repair time), After taking Laplace transformation we get, get $B^*(\lambda)=\left(\dfrac{r\mu}{r\mu+\lambda}\right)^r$, $B^*(\lambda+\alpha)=\left(\dfrac{r\mu}{r\mu+\lambda+\alpha}\right)^r$, $B^*(\lambda+2\alpha)=\left(\dfrac{r\mu}{r\mu+\lambda+2\alpha}\right)^r$.
   By (equation (38)), where Π_2, Ψ_2 & Ω_2 have the usual meaning as before for the gamma repair time distribution.

$$A_{V(Gam(r))} = \left[1+\dfrac{(\lambda+3\alpha)\{1+q(1-q)\}\{(r\mu+\lambda+2\alpha)^r-(r\mu)^r\}}{\mu(\lambda+2\alpha)(r\mu)^r}+\left(2-q+\dfrac{\alpha}{\lambda}\right)\psi_2+\dfrac{1}{\lambda}\{\lambda(1-q)q^2+\Omega_2-\Pi_2\}\right]P_{1,3} \tag{40}$$

III. **Uniform distribution:** In this scenario, repair time is distributed uniformly over the time span $[a, b]$. We set $b_1 = (a + b)/2$ (mean repair time). We obtain the following values after taking the Laplace transformation. $B^*(\lambda) = \dfrac{e^{-a\lambda} - e^{-b\lambda}}{\lambda(b-a)}$,

$$B^*(\lambda+\alpha) = \frac{e^{-a(\lambda+\alpha)} - e^{-b(\lambda+\alpha)}}{(\lambda+\alpha)(b-a)}, \ B^*(\lambda+2\alpha) = \frac{e^{-a(\lambda+2\alpha)} - e^{-b(\lambda+2\alpha)}}{(\lambda+2\alpha)(b-a)}$$

By equation (38), where Π_3, Ψ_3 & Ω_3 have the usual meaning as before for the uniform repair time distribution.

$$A_{V[U(a,b)]} = \left[1 + \frac{(\lambda+3\alpha)\{1+q(1-q)\}\{(\lambda+2\alpha)(b-a)-\left(e^{-a(\lambda+2\alpha)} - e^{-b(\lambda+2\alpha)}\right)\}}{(\lambda+2\alpha)\left(e^{-a(\lambda+2\alpha)} - e^{-b(\lambda+2\alpha)}\right)} + \left(2-q+\frac{\alpha}{\lambda}\right)\psi_3 + \frac{1}{\lambda}\left\{\lambda(1-q)q^2 + \Omega_3 - \Pi_3\right\}\right]P_{1,3} \ (41)$$

5. Comparison of availability

In this study, the comparison of the availability of three types of repair time distributions such as exponential $A_{V(M)}$, gamma $A_{VGamma(2.5)}$, and uniform $A_{VU[2, 18]}$ are used. The values of the different parameters are set as follows.

$$\frac{1}{\lambda} = 2000 \text{ days}; \ \frac{1}{\alpha} = 5000 \text{ days}; \ \frac{1}{\mu} = 10 \text{ days}; \ \frac{1}{\beta} = \frac{10}{24} \text{ days}; \ \frac{1}{q} = 20 \text{ days} \quad \text{i.e. } \lambda =$$

0.0005; $\alpha = 0.0002$; $\mu = 0.1$; $\beta = 2.4$; $q = 0.05$ [Ref. 2].

We consider three cases for numerical computation that are given in the following Tables 2.1–2.3.

Table 2.1 The value of λ vary from 0.0005 to 0.005 and values of α, μ, β, q are kept fixed.

λ	α	β	μ	q	$A_{V(M)}$	$A_{VGamma(2.5)}$	$A_{VU[2, 18]}$
0.0005	0.0002	2.4	0.1	0.05	0.998960239	0.998961175	0.998961493
0.001	0.0002	2.4	0.1	0.05	0.997625831	0.9976263	0.997626051
0.0015	0.0002	2.4	0.1	0.05	0.994914519	0.994904536	0.994901742
0.002	0.0002	2.4	0.1	0.05	0.993830216	0.993804998	0.993794441
0.0025	0.0002	2.4	0.1	0.05	0.991398708	0.991336045	0.991315266
0.003	0.0002	2.4	0.1	0.05	0.988628450	0.988507067	0.988468512
0.0035	0.0002	2.4	0.1	0.05	0.985531543	0.985340277	0.985266882
0.004	0.0002	2.4	0.1	0.05	0.982108682	0.981806674	0.981709079
0.0045	0.0002	2.4	0.1	0.05	0.97837562	0.977950852	0.977808623
0.005	0.0002	2.4	0.1	0.05	0.974364347	0.973778886	0.9735602

Table 2.2 The value of β vary from 1 to 10 and values of λ, α, μ, q are kept fixed.

λ	α	β	μ	q	$A_{V(M)}$	$A_{VGamma(2.5)}$	$A_{VU[2, 18]}$
0.0005	0.0002	1	0.1	0.05	0.998380558	0.99838149	0.99838189
0.0005	0.0002	2	0.1	0.05	0.998877235	0.998878420	0.998878857
0.0005	0.0002	3	0.1	0.05	0.999042903	0.999044036	0.99904485

λ	α	β	μ	q	$A_{V(M)}$	$A_{VGamma(2.5)}$	$A_{VU[2,18]}$
0.0005	0.0002	4	0.1	0.05	0.999125893	0.999126895	0.999127406
0.0005	0.0002	5	0.1	0.05	0.999175437	0.999176720	0.999176886
0.0005	0.0002	6	0.1	0.05	0.999208155	0.9992099	0.999210227
0.0005	0.0002	7	0.1	0.05	0.999232551	0.999233567	0.999233654
0.0005	0.0002	8	0.1	0.05	0.999250067	0.999251268	0.99925123
0.0005	0.0002	9	0.1	0.05	0.999263324	0.999265076	0.999265389
0.0005	0.0002	10	0.1	0.05	0.999274859	0.999276425	0.999276553

Table 2.3 The value of q vary from 0.01 to 0.1 and values of λ, α, μ, β are kept fixed.

λ	α	β	μ	q	$A_{V(M)}$	$A_{VGamma(2.5)}$	$A_{VU[2,18]}$
0.0005	0.0002	2.4	0.1	0.01	0.998993631	0.99899120	0.998990346
0.0005	0.0002	2.4	0.1	0.02	0.998987568	0.998986363	0.998986009
0.0005	0.0002	2.4	0.1	0.03	0.998980627	0.998979841	0.998979691
0.0005	0.0002	2.4	0.1	0.04	0.998971175	0.998971587	0.998971898
0.0005	0.0002	2.4	0.1	0.05	0.998960002	0.998961225	0.998961652
0.0005	0.0002	2.4	0.1	0.06	0.998947798	0.998948987	0.998949425
0.0005	0.0002	2.4	0.1	0.07	0.998932625	0.998934454	0.998935958
0.0005	0.0002	2.4	0.1	0.08	0.998915562	0.998918982	0.998920775
0.0005	0.0002	2.4	0.1	0.09	0.998896639	0.998901486	0.998902638
0.0005	0.0002	2.4	0.1	0.1	0.998875987	0.998881346	0.998883819

6. Interpretation of the result

This system is made up of one primary and three standby components that deal with reboot delays and general repair times. When the primary component fails, a standby component takes its place unless all standby components are exhausted. The failed units are repaired on FCFS manner. First, system's availability is estimated by the supplemental variable approach and LST (Laplace-Stieltjes transformation). We examine numerical findings based on explicit representations of availability designed for three distinct distributions of repair time, namely exponential, gamma, and uniform.

References

1.   Cox, D. R. (1955). The analysis of non-Markovian stochastic processes by the inclusion of supplementary variables. Math. Proc. Cambridge Phil. Soc., 51(3), 433–441.
2.   Wang, K. H. and Chen, Y. J. (2009). Comparative analysis of availability between three systems with general repair times, reboot delay and switching failures. App. Math., 215(1), 384–394.
3.   Yen, T. C., Chen, W. L., and Wang K. H. (2020). Comparison of three availability system with warm standby components, detection delay and general repair times. Symmetry, 12(3), 414.

3 A study of non-Newtonian K-L fluid model through an inclined artery of non-uniform cross-section

Amendra Singh[a] and Sanjeev Kumar[b]

Department of Mathematics, IBS, Khandari Campus, Dr. Bhimrao Ambedkar University, Agra-282002, Uttar Pradesh, India

Abstract

This work investigates the impact of blood flow parameters within an inclined artery featuring an irregular cross-sectional shape and multiple constrictions. In this study, blood is modeled using K-L fluid model, with no-slip boundary conditions applied to the arterial wall. Numerical expressions have been derived for the flow characteristic. The numerical expressions have been resolved using the Gauss-Kronrod quadrature formula in Matlab R2022b. The effects of various flow parameters namely stenoses heights, yield stress, angle of inclination, wall exponent parameter and plasma viscosity on flow characteristics have been studied and discussed graphically. The findings indicate that the flow characteristic are notably influenced by the non-Newtonian (K-L) fluid. Further, we have also observed that the inclination angle and non-uniform cross-section significantly affect the flow pattern, with the flow becoming more complex as the inclination angle and non-uniformity increase. The outcomes of the work have been verified by other models which already exist.

Keywords: Stenosis, blood flow rate, resistance to blood flow, skin fraction, non-uniform channel

Introduction

Throughout history, researchers have conducted various studies to comprehend the movement of blood within the cardiovascular system. Among the topics of discussion is the issue of blood flow in arteries that have stenoses. Researchers have examined the behavior of blood in these situations, considering both Newtonian as well as non-Newtonian fluids have been used to analyze blood quantities. In 1968, Young introduced the concept of time-dependent arteries and analyzed the effects on fluid flow through a tube. He proposed an approximate solution for the time-dependent mild stenosis problem. Kumar and Kumar (2006) investigated axisymmetric blood circulation within inflexible constricted tube and also (Kumar et al., 2009) analyze the effect of MHD flow in a stenosed artery. Awgichew et al. (2013) considered a non-uniform channel for fluid flow and analyzed various flow parameters such as Darcy number and wall exponent parameter.

Numerous researchers have explored various non-Newtonian models like Kuang-Luo, Power-law, Herschel-Bulkley, and Casson models in recent times, yielding significant findings in the domain of cardiovascular disorders. Bali and

[a]amendra1729@gmail.com, [b]sanjeevibs@yahoo.co.in

Gupta (2018) examined the blood circulation in a non-symmetric stenosed artery. Kumar et al. (2020) conducted a comparative analysis of non-Newtonian blood circulation within a stenosed artery with elasticity. Prasad et al. (2020), focused on assessing the impact of an inclined angle on blood circulation within a ste-nosed artery, in this investigation blood is modeled using a power-law rheological model. They found significant results on flow characteristics. Owasit and Sriyab (2021) formulated a mathematical model to investigate blood flow through mul-tiple stenoses, specifically the bell and cosine combined shape of stenoses in a two-dimensional artery, and verified these outcomes with other theoretical mod-els. Singh and Kumar (2023) considered Newtonian fluid in an inclined artery and predict blood flow parameters in the presence of multiple constrictions.

Inspired by the aforementioned research, it has come to light that numerous veins undergo gradual changes in their cross-section along their length, with many exhibiting multiple stenoses. In the realm of physiological systems, there exist several blood vessels that incline with the axis. Thus, we have developed a mathematical model to investigate the influence of flow parameters on the flow characteristic in an irregular inclined channel with two constrictions. Through this, we have uncovered the influence of numerous blood flow parameters on key flow characteristics, including blood flow rate, resistance to blood flow and skin friction. This mathematical model can offer valuable insights into the role of vari-ous flow parameters in addressing blood flow issues.

Mathematical formulation

Consider an axially symmetric inclined artery having an angle α with the z-axis, with cylindrical polar coordinates (r, θ, z) and assumed non-uniform rigid tube of radius $R(z)$. As discussed, the blood is taken as non-Newtonian (K-L) fluid with constant density and viscosity and blood flow is taken as laminar and steady. Further, considering the radius of stenosis, uniform and non-uniform region in an artery is $R(z)$, R_0 and R_1 respectively. δ_1 and δ_2 are the height of the first and second stenosis respectively. d_1 and L_0 are the lengths of the first non-stenosis segment and the uniform rigid tube respectively. l_1 and l_2 are the lengths of the first and second stenosis segments. L is considered the total length of the arterial segment.

The geometry of the problem shown in Figure 3.1 and described in the equa-tion (1)

$$R/R_0 = \begin{cases} 1 - \frac{\delta_1}{2R_0}\left[1 + cos\frac{2\pi}{l_1}\left(z - d_1 - \frac{l_1}{2}\right)\right]; & d_1 \leq z \leq d_1 + l_1 \\ 1 - \frac{\delta_2}{2R_0}\left[1 + cos\frac{2\pi}{l_2}(z - L_0)\right]; & L_0 - \frac{l_2}{2} \leq z \leq L_0 \\ R_1 - \frac{\delta_2}{2R_0}\left[1 + cos\frac{2\pi}{l_2}(z - L_0)\right]; & L_0 \leq z \leq L_0 + \frac{l_2}{2} \\ R_1; & L_0 + \frac{l_2}{2} \leq z \leq L \\ 1; & otherwise \end{cases} \tag{1}$$

Further, R_1 is expressed as $\frac{1}{R_0}e^{\beta L^2(z-L_0)^2}$ (Awgichew et al., 2013)

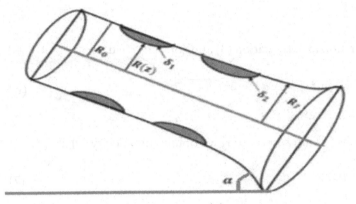

Figure 3.1 Blood flow geometry of the artery

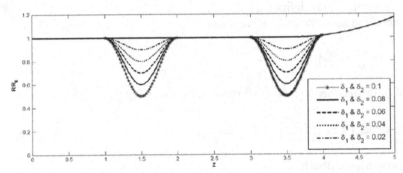

Figure 3.2 Stenosed geometry for different values of stenoses height δ_1 and δ_2

The governing equation (Prashad et al., 2008) of the flow is

$$\frac{1}{r}\frac{\partial}{\partial r}(r\tau) = -\frac{\partial p}{\partial z} + \frac{sin\alpha}{F} \tag{2}$$

where $F = \frac{\mu C}{\rho g R_0^2}$ and τ is the skin friction of K-L fluid [6]

$$\tau = \tau_y + \eta_2\dot{\gamma}^{1/2} + \eta_1\dot{\gamma} \qquad \tau > \tau_y$$
$$\tag{3}$$
$$\dot{\gamma} = 0 \qquad \tau < \tau_y$$

where τ_y, $\dot{\gamma}$, η_1, η_2 are yield stress, shear rate, plasma viscosity and other chemical substance, respectively.

The boundary conditions are:

$$\left.\begin{array}{l} \tau \text{ is finite at } r = 0 \\ v = 0 \text{ at } r = R_0 \text{ and } R(z) \end{array}\right\} \tag{4}$$

Solution to the problem

Solving momentum equation for shear stress, we get

$$\tau = \frac{(P+\theta)}{2}r \tag{5}$$

Substitute τ from equation (5) in equation (3) and solve for yield stress $\dot{\gamma} = \left(-\frac{\partial v}{\partial r}\right)$

$$\left(-\frac{\partial v}{\partial r}\right) = \dot{\gamma} = \frac{1}{4\eta_1{}^2}\left(\eta_2{}^2 + (A + Br) - 2\eta_2\sqrt{(A + Br)}\right) \tag{6}$$

where $A = \eta_2{}^2 - 4\eta_1\tau_y$, $B = 2\eta_1(P + \theta)$, $P = -\frac{\partial p}{\partial z}$ and $\theta = \frac{\sin\alpha}{F}$

Integrating equation (6) with respect to r, we obtain the velocity (v) as

$$v = \frac{1}{4\eta_1{}^2 B}(C + BDr) \tag{7}$$

Where $C = \frac{1}{2}A^2 + \eta_2^2 A - \frac{4}{3}\eta_2 A^{\frac{3}{2}}$ and $D = A + \eta_2^2 - 2\eta_2 A^{\frac{1}{2}}$

Volumetric flow rate (Q) is defined as: $Q = 2\int_{r=0}^{r=R} v\,r\,dr$
Using equation (7) in above definition, we obtain

$$Q_S = \left[\frac{1}{2\eta_1{}^2}\left(\frac{CR^2}{2B} + \frac{DR^3}{3}\right)\right] \tag{8}$$

The ratio of Q in stenosed artery (Q_S) to normal artery (Q_N) is

$$Q = \frac{Q_S}{Q_N} = \left[\frac{1}{2\eta_1{}^2}\left(\frac{CR^2}{2B} + \frac{DR^3}{3}\right)\right] \times \left[\frac{1}{2\eta_1{}^2}\left(\frac{CR_0^2}{2B} + \frac{DR_0^3}{3}\right)\right]^{-1} \tag{9}$$

From equation (8), we obtain

$$P = \frac{CR^2}{4\eta_1}\left(\frac{3}{6\eta_1^2 Q_S - DR^3}\right) - \theta \tag{10}$$

Pressure drop per wave length is $\Delta p = \int_0^L \left(-\frac{\partial p}{\partial z}\right)dz$

$$\Delta p = \int_0^L \left(\frac{CR^2}{4\eta_1}\left(\frac{3}{6\eta_1^2 Q_S - DR^3}\right) - \theta\right)dz \tag{11}$$

Resistance to flow is $\lambda = \frac{\Delta p}{Q}$

Using equation (11) in the above definition, we obtain

$$\lambda_S = \left[\frac{1}{Q}\int_0^L \left(\frac{CR^2}{4\eta_1}\left(\frac{3}{6\eta_1^2 Q_S - DR^3}\right) - \theta\right)dz\right] \tag{12}$$

The ratio of λ in stenosed artery (λ_S) to normal artery $(\lambda_N)$is

$$\lambda = \frac{\lambda_S}{\lambda_N} = \left[\frac{1}{Q}\int_0^L \left(\frac{CR^2}{4\eta_1}\left(\frac{3}{6\eta_1^2 Q_S - DR^3}\right) - \theta\right)dz\right] \times \left[\frac{1}{Q}\int_0^L \left(\frac{CR_0^2}{4\eta_1}\left(\frac{3}{6\eta_1^2 Q_S - DR_0^3}\right) - \theta\right)dz\right]^{-1} \tag{13}$$

Skin friction is $\tau = -\frac{R}{2}\frac{\partial p}{\partial z}$

Using equation (10) in the above definition, we obtain

$$\tau_S = \left[\frac{CR^2}{8\eta_1}\left(\frac{3}{6\eta_1^2 Q_S - DR^3}\right) - \theta\right] \tag{14}$$

The ratio of τ in stenosed artery (τ_S) to normal artery (τ_N) is

$$\tau = \frac{\tau_S}{\tau_N} = \left[\frac{CR^2}{8\eta_1}\left(\frac{3}{6\eta_1^2 Q_S - DR^3}\right) - \theta\right] \times \left[\frac{CR_0^2}{8\eta_1}\left(\frac{3}{6\eta_1^2 Q_S - DR_0^2}\right) - \theta\right]^{-1} \tag{15}$$

All integrations in equation (13) are solved numerically through the Gauss-Kronrod quadrature formula by using Matlab R2022b.

Numerical results and discussions

In this section, we analyze the volumetric blood flow rate (Q), resistance to blood flow (λ) and skin friction (τ) through given equations (9, 13, 15). These parameters are also evaluated numerically by using Matlab R2022b software for the various parametric values provided in Table 3.1. The graphical representation of the flow parameters demonstrates their respective effects.

Table 3.1 Numerical values of parameters.

Parameters	Numerical values
Blood flow (Q)	$0 - 5$ cm^3/$_S$
Artery length (l)	5 cm
Inclined artery angle (α)	$0 - 60°$
Radius of the artery (R_0)	0.02 cm
Wall exponent parameter (k)	$0 - 0.5$
Stenosis height (δ)	$0 - 0.01$ cm

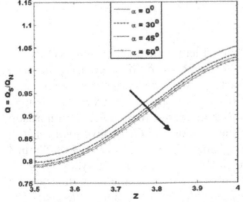

Figure 3.3 Variation in Q with z for α ($\tau_y = 2$, $\eta_1 = 3$, $\delta_1 = 0.01$, $\delta_2 = 0.02$, $k = 0.1$)

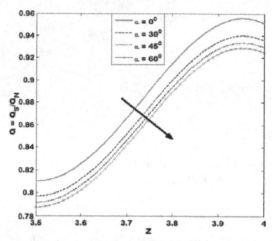

Figure 3.4 Variation in Q with z for α ($\tau_y = 2$, $\eta_1 = 3$, $\delta_1 = 0.01$, $\delta_2 = 0.02$, $k = -0.1$)

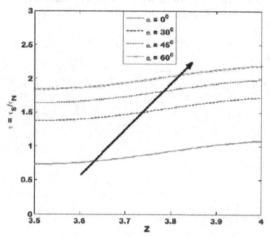

Figure 3.5 Variation in τ with z for α ($\tau_y = 2$, $\eta_1 = 3$, $\delta_1 = 0.01$, $\delta_2 = 0.02$, $k = 0.1$)

Figure 3.2 illustrates the geometry of the problem with varying stenosis heights. The observation reveals that as the height of stenoses (δ_1, δ_2) increases, the ratio of the stenosed artery's radius to normal artery (R/R_0) decreases. Figures 3.3 and 3.4 depict the changes in Q as a function of axial distance (z) for various inclined angle (α). It is observed that Q decreases as α increases. Moreover, a comparison between Figures 3.3 and 3.4 reveals that the variation in Q for given values ($\tau_y = 2$, $\eta_1 = 3$, $\delta_1 = 0.01$, $\delta_2 = 0.02$) is more pronounced in a divergent ($k > 0$) channel compared to a convergent ($k < 0$) channel at the non-uniform artery.

Figures 3.5 and 3.6 depict the variation of τ with the angle of the inclined artery(α). It is observed that τ increases as α increases, while it decreases as plasma viscosity (η_1) increases. The variation in λ is shown in Figures 3.7–3.10

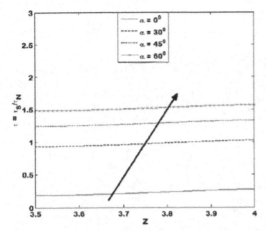

Figure 3.6 Variation in τ with z for α ($\tau_y = 2$, $\eta_1 = 5$, $\delta_1 = 0.01$, $\delta_2 = 0.02$, $k = 0.1$)

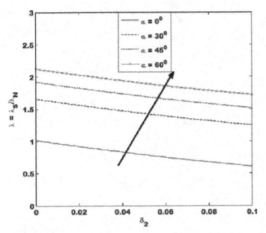

Figure 3.7 Variation in λ with δ_2 for α ($\tau_y = 2$, $\eta_1 = 3$, $\delta_1 = 0.01$, $k = 0.1$)

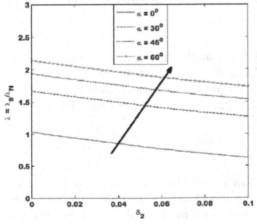

Figure 3.8 Variation in λ with δ_2 for α ($\tau_y = 2$, $\eta_1 = 3$, $\delta_1 = 0.01$, $k = 0.3$)

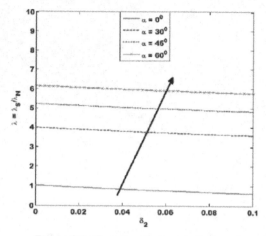

Figure 3.9 Variation in λ with δ_2 for α ($\tau_y = 2$, $\eta_1 = 5$, $\delta_1 = 0.01$, $k = 0.1$)

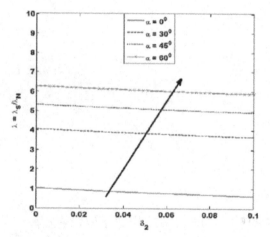

Figure 3.10 Variation in λ with δ_2 for α ($\tau_y = 4$, $\eta_1 = 5$, $\delta_1 = 0.01$, $k = 0.1$)

for the different values of the inclined artery angle (α). Figures 3.7 and 3.8 illustrate that λ increases as the angle of the inclined artery (α) increases. However, it increases very slightly as the wall exponent parameter (k) increases. Furthermore, it is observed that there is no significant impact (Figures 3.9 and 3.10) of yield stress (τ_y) on λ.

Conclusion

In this investigation, we analyze the flow of blood within an inclined artery of a non-uniform channel, considering the blood as a K-L fluid. The computed results for volumetric blood flow rate, resistance to blood flow and skin friction were visually represented for numerous parameters such as inclined angle, yield stress, plasma viscosity, stenosis height and wall exponent parameter. The results show that the plasma viscosity significantly affects all flow quantities, whereas yield stress has no significant changes. The inclination angle and non-uniform cross-section also affect the flow pattern and it becomes more complex as the inclination angle and wall exponent parameter increase. Further, the volumetric flow rate is more for a divergent channel ($k > 0$) as compared to a convergent channel ($k < 0$) at the end of the non-uniform artery. This study has important implications for the diagnosis and treatment of cardiovascular diseases.

References

1. Young, D. F. (1968). Effect of a time-dependent stenosis on flow through a tube. *J. Engg. Ind.*, 90(2), 248–254.
2. Kumar, S. and Kumar, S. (2006). Numerical study of axisymmetric blood flow in a constricted rigid tube. *Int. Rev. Pure Appl. Math.*, 2(2), 99–109.
3. Prashad, K. M. and Radhakrishnamacharya, G. (2008). Flow of Harschel-Bulkley fluid through an inclined tube of non-uniform cross section with multiple stenosis. *Arch. Mech.* 60, 161–172.
4. Kumar, S., Kumar, S., and Kumar, D. (2009). Oscillatory MHD flow of blood through an artery with mild stenosis. *Int. J. Engg. Trans. A: Basic*, 22(2), 125–130.
5. Awgichew, G. and Radhakrishnamacharya, G. (2013). Effects of slip conditions and multiple constrictions on couple stress fluid flow through a channel of non-uniform cross section. *ARPN J. Engg. Appl. Sci.*, 8(5), 393–400.
6. Bali, R. and Gupta, N. (2018). Study of non-Newtonian fluid by K-L model through a non-symmetrical stenosed narrow artery. *Appl. Math. Comput.*, 320(2018), 358–370.
7. Kumar, S., Kumar S., and Kumar, D. (2020). Comparative study of non-Newtonian physiological blood flow through elastic stenotic artery with rigid body stenotic artery. *Series Biomec.*, 34(4), 26–41.
8. Prasad, K. M. and Yasa, P. R. (2020). Flow of non-Newtonian fluid through a permeable artery having non-uniform cross section with multiple stenosis. *J. Naval Arch. Marine Engg.*, 17(1), 31–38.
9. Owasit, P. and Sriyab S. (2021). Mathematical modeling of non-Newtonian fluid in arterial blood flow through various stenoses. *Adv. Diff. Eqn.*, 340(2021), 1–20.
10. Singh, A. and Kumar, S. (2023). Prediction of blood flow parameters in an inclined artery with multiple stenoses. *Series Biomec.*, 37(2), 74–84.

4 Non-split domination cover pebbling number for some class of middle graphs

A. Lourdusamy[1,a], I. Dhivviyanandam[2,b] and Lian Mathew[3,c]

[1,2]Department of Mathematics, St. Xavier's College (Autonomous), Palayamkottai, Affiliated to Manonmaniam Sundaranar University, Abisekapatti-627012, Tamil Nadu, India

[3]Department of Mathematics, CHRIST (Deemed to be University), Lavasa Campus, Pune, India

Abstract

A pebbling shift is the process of removing a pair of pebbles from a vertex and keeps one pebble at a neighboring vertex and eliminating the other for the recurring cost. The non-split domination cover pebbling number $\Psi_{ns}(G)$, is the least amount of pebbles needed on the vertices of the graph G, such that the set of vertices with at least one pebble becomes a non-split dominating set of G after a series of pebbling move for all distributions of pebbles on G. We discuss non-split domination number and determine Ψ_{ns} for some families of middle graphs.

Keywords: Dominating set, cover pebbling number, domination cover pebbling, middle graphs
2020 AMS Subject Classification: 05C12, 05C25.

Introduction

Lagarias and Saks were the first ones to introduce the concept of pebbling and F. R. K. Chung (1989) used the concept of pebbling to solve a number theoretic conjecture. Then many authors started studying pebbling concepts. Glenn Hulbert (1999) wrote a survey of pebbling variants. The subject of graph pebbling has seen massive growth after Hulbert's survey. In the past 30 years, so many new variants in graph pebbling have been developed which can be applied to the fields of transportation, computer memory allocation, game theory, and the installation of mobile towers.

Let us denote G's vertex and edge sets as V(G) and E(G), respectively. Consider a graph with a fixed number of pebbles at each vertex. One pebble is thrown away and the other is placed on an adjacent vertex when two pebbles are removed from a vertex. This process is known as a pebble move. The pebbling number of a vertex v in a graph G is the smallest number $\pi(G, v)$, allowing us to shift a pebble to v using a sequence of pebbling moves, regardless of how these pebbles are located on G's vertices. The pebbling number, $\pi(G)$, of a graph G is the maximum of $\pi(G, v)$ over all the vertices v of a graph. Considering the concepts of cover pebbling

---

[a]lourdusamy15@gmail.com, [b]divyanasj@gmail.com, [c]lianmathew64@gmail.com

and non-split domination, we develop the concept: non-split domination cover pebbling number of a graph, denoted by $\Psi_{ns}(G)$. Crull et al. (2005) defined the cover pebbling number as: "The cover pebbling number, $\lambda(G)$ is defined as the minimum number of pebbles required such that given any initial configuration of at least $\lambda(G)$ pebbles, it is possible to make a series of pebbling moves to place at least one pebble on every vertex of G". Crull et al. (2005) defined the domination cover pebbling number as: The domination cover pebbling number is defined as "the minimum number of pebbles required so that any initial configuration of pebbles can be shifted by a sequence of pebbling moves so that the set of vertices that contain pebbles form a dominating set S of G". Kulli et al. (2000) introduced the non-split domination number. "A dominating set D of a graph G = (V, E) is a non-split dominating set if the induced *graph* <V–D> is connected. The non-split domination number $\gamma_{ns}(G)$ of G is the minimum cardinality of a non-split domi-nating set". The non-split domination cover pebbling number, $\Psi_{ns}(G)$, is the least amount of pebbles that must be placed on the vertices of the graph G in order for the set of vertices with at least one pebble to become a non-split dominating set of G after a series of pebbling move for all distributions of pebbles on G. We denote the non-split domination set as NSDS. We discuss non-split domination pebbling number (NSDPN) and determine Ψ_{ns} for some families of middle graphs.

Preliminaries

For graph-theoretic terminology, we refer to J. A. Bondy and U. S. R. Murty (1977), Gary Chartrand (1985).

Results

1. For a simple connected graph G, $\Psi(G) \leq \Psi_{ns}(G) \leq \sigma(G)$.
2. For K_n, $\Psi_{ns}(K_n) = 1$.
3. For W_n, | $\Psi_{ns}(W_n) = \Psi(W_n)$.

NSDPN for the middle graph of path
Theorem 3.1.1: Consider the middle graph of the path. Then, $\Psi_{ns}(M(P_n)) = 2^{n+1} - 3$.

Proof: Let $V(P_n) = \{x_1, x_2, x_3, ..., x_n\}$ and $y_1, y_2, ..., y_{n-1}$ be the inserted vertices between the edges $e_1, e_2, ..., e_{n-1}$ of P_n to construct $M(P_n)$. Then the total number of vertices is $2n - 1$ and the edges are $3n - 4$. Consider the non-split domination set $D = \{x_i \mid 1 \leq i \leq n\}$ with containing n vertices. The vertices of D dominate all the vertices in $M(P_n)$ and <V – D> is connected. Placing $2^{n+1} - 4$ pebbles on any one of the end vertices we could cover the non-split dominating set's vertices x_n to x_2, if we place all the pebbles at x_n. And also, we don't get the non-split domination set. Hence, the NSDPN of $M(P_n)$ is $\Psi_{ns}(M(P_n)) \geq 2^{n+1} - 3$.

Considering the configuration of C with $2^{n+1} - 3$ on $V(M(P_n))$, we prove $\Psi_{ns}(M(P_n)) \leq 2^{n+1} - 3$ to cover D.

Case 1: Let the source vertex be x_1.

Let $p(x_1) = 2^{n+1} - 3$. To cover x_n we require only 2^n pebbles. Thus, to cover the next furthest non-split dominating set is x_{n-1} which requires 2^{n-1}. Likewise, we get the series of pebble distribution $2^n + 2^{n-1} + 2^{n-2} + \ldots + 2^2 + 2^0$ to cover all the vertices of D. Thus, we used $\sum_{k=2}^{n} 2^k + 1 = 2^{n+1} - 3$ pebbles.

Case 2: Let the source vertex be x_l, $1 < l \le n - 1$.
Let us place all the pebbles at x_1. If $< \lfloor \frac{n}{2} \rfloor$, then we need $2^{(n-l)+1} - 3$ pebbles to cover x_l to x_n of the non-split dominating set. Now to cover D we need $2^{l+1} - 4$. Thus, we used $2^{(n-l)+1} + 2^{l+1} - 7 < 2^{n+1} - 3$ pebbles.

Case 3: Let the source vertex be y_1.
Let us consider the source vertex either any one of the non-split dominating vertices of y_1 or $y_n - 1$. Consider all the pebbles placed on y_1. Then we need 4 pebbles to cover the adjacent vertices x_1 and x_2. Then to cover the remaining NSDS of vertices we need $2^n - 4$ pebbles. Thus, we used $2^n < 2^{n+1} - 3$ pebbles. Hence, $\Psi_{ns}(M(P_n)) = 2^{n+1} - 3$.

NSDPN *for the middle graph of cycle graphs*
Theorem 3.2.1: Consider the middle graph of the cycle. Then, $\Psi_{ns}(M(C_n)) =$

$$\begin{cases} 2\sum_{k=0}^{\lfloor \frac{n}{2} \rfloor} 2^k - 8, & n \text{ is odd} \\ \sum_{k=1}^{\lfloor \frac{n}{2} \rfloor + 1} 2^k - 8 + \sum_{k=1}^{\lfloor \frac{n}{2} \rfloor} 2^k, & n \text{ is even} \end{cases}$$

Proof: Let $V(C_n) = \{x_1, x_2, x_3, \ldots, x_n\}$ and $y_1, y_2, \ldots, y_n$ be the inserted vertices between the edges $e_1, e_2, \ldots, e_n$ of C_n to construct the middle graph of cycle $M(C_n)$. Then the total number of vertices is $2n$ and the edges are $3n$. Let the non-split dominating set $D = \{y_i \cup x_1, x_2, x_3, \ldots, x_n\}$ where $1 \le i, j \le n$ and $x_j \ne N[y_j]$. Thus, D dominates all the vertices of $M(C_n)$, and $<V - D>$ is connected. The total number of vertices in D is $n - 1$.

Case 1: When n is odd.
Without loss of generality, Let $D = \{y_n, x_2, x_3, \ldots, x_{n-1}\}$. Placing $2\sum_{k=0}^{\lfloor \frac{n}{2} \rfloor} 2^k - 7$ pebbles on the source vertex x_1 we cannot put one pebble each on all the vertices of D. Hence, $\psi_{ns}(M(C_n)) \ge 2\sum_{k=0}^{\lfloor \frac{n}{2} \rfloor} 2^k - 8$ when n is odd.

Distribution of $2\sum_{k=0}^{\lfloor \frac{n}{2} \rfloor} 2^k - 8$ pebbles on the configuration of C, we prove $\psi_{ns} \le 2\sum_{k=0}^{\lfloor \frac{n}{2} \rfloor} 2^k - 8$ to cover D.
Case 1.1: Let the source vertex be x_1.
Using **Theorem 3.1.1**, we can cover the non-split dominating set of D from x_2 to $x_{\lfloor \frac{n}{2} \rfloor}$. Thus, $2^{\lfloor \frac{n}{2} \rfloor + 1} - 4$ pebbles are used to cover D. Similarly, to cover the remaining NSDS $\{y_n, x_{n-1}, \cdots, x_{\lfloor \frac{n}{2} \rfloor + 1}\}$ we use $2^{\lfloor \frac{n}{2} \rfloor + 1} - 6$ pebbles. Hence, we have spent total $2^{\lfloor \frac{n}{2} \rfloor + 2} - 10 = 2\sum_{k=0}^{\lfloor \frac{n}{2} \rfloor} 2^k - 8$ pebbles.
Case 1.2: Let the source vertex be any one of the vertices of y_i, $1 \le i \le n$.

Without loss of generality, let the source vertex be y_n. Using the **Theorem 3.1.1**, we can cover the non-split dominating set of D from x_2 to $x_{\lceil\frac{n}{2}\rceil}$. Thus, $2^{\lceil\frac{n}{2}\rceil+1} - 4$ pebbles are used to cover D. Similarly, to cover the remaining non-split domi-nating set $\{y_n, x_{n-1}, \cdots, x_{\lceil\frac{n}{2}\rceil+1}\}$ we use $2^{\lceil\frac{n}{2}\rceil} - 3$ pebbles. Hence, we have spent total $3\left(2^{\lceil\frac{n}{2}\rceil}\right) - 7 < 2\sum_{k=0}^{\lceil\frac{n}{2}\rceil} 2^k - 8$ pebbles. Hence, $\psi_{ns}(M(C_n)) \leq 2\sum_{k=0}^{\lceil\frac{n}{2}\rceil} 2^k - 8$, When n is odd.

Case 2: When n is even.

Without loss of generality, Let $D = \{y_n, x_2, x_3, \ldots, x_{n-1}\}$. Placing $\sum_{k=1}^{\lceil\frac{n}{2}\rceil+1} 2^k - 7 + \sum_{k=1}^{\lceil\frac{n}{2}\rceil} 2^k$ pebbles on the source vertex x_1, we cannot put one pebble each on all the vertices of D. Hence, $\psi_{ns}(M(C_n)) \geq \sum_{k=1}^{\lceil\frac{n}{2}\rceil+1} 2^k - 8 + \sum_{k=1}^{\lceil\frac{n}{2}\rceil} 2^k$, When n is even.

Distribution of $\sum_{k=1}^{\lceil\frac{n}{2}\rceil+1} 2^k - 8 + \sum^{\lceil\frac{n}{2}\rceil} 2^k$ pebbles on the configuration of C, we cover all the vertices of D. Now we prove the sufficient condition for $M(C_n)$, when n is even.

Case 2.1: Let the source vertex be x_1.

Using the **Theorem 3.1.1**, we can cover the non-split dominating set of D from x_2 to $x_{\lceil\frac{n}{2}\rceil+1}$. Thus, $2^{\lceil\frac{n}{2}\rceil+2} - 44$ pebbles are used to cover D. Similarly, to cover the remaining non-split dominating set $\{y_n, x_{n-1}, \cdots, x_{\lceil\frac{n}{2}\rceil+2}\}$ we use $2^{\lceil\frac{n}{2}\rceil+1} - 6$ pebbles. Hence, we have spent total $3\left(2^{\lceil\frac{n}{2}\rceil+1}\right) - 10 = \sum_{k=1}^{\lceil\frac{n}{2}\rceil+1} 2^k - 8 + \sum_{k=1}^{\lceil\frac{n}{2}\rceil} 2^k$ pebbles.

Case 2.2: Let the source vertex be any one of the vertices of y_i, $1 \leq i \leq n$.

Without loss of generality, let the source vertex be y_n. Using the **Theorem 3.1.1**, we can cover the non-split dominating set of D from x_2 to $x_{\lceil\frac{n}{2}\rceil+1}$. Thus, $2^{\lceil\frac{n}{2}\rceil+2} - 4$ pebbles are used to cover D. Similarly, to cover the remaining non-split domi-nating set $\{y_n, x_{n-1}, \cdots, x_{\lceil\frac{n}{2}\rceil+2}\}$ we use $2^{\lceil\frac{n}{2}\rceil} - 3$ pebbles. Hence, we have spent total $5\left(2^{\lceil\frac{n}{2}\rceil}\right) - 7 < \sum_{k=1}^{\lceil\frac{n}{2}\rceil+1} 2^k - 8 + \sum_{k=1}^{\lceil\frac{n}{2}\rceil} 2^k$ pebbles. Hence, $\psi_{ns}(M(C_n)) \leq \sum_{k=1}^{\lceil\frac{n}{2}\rceil+1} 2^k - 8 + \sum_{k=1}^{\lceil\frac{n}{2}\rceil} 2^k$, when n is even. Hence proved.

NSDPN for the middle graph of wheel graphs
Theorem 3.3.1: Consider the middle graph of the wheel. Then,

$$\psi_{ns}(M(W_n)) = \begin{cases} \lceil\frac{n}{2}\rceil 8 + 6, & n \text{ is odd} \\ \lceil\frac{n}{2}\rceil 8 + 10, & n \text{ is even.} \end{cases}$$

Proof: Let $V(W_n) = \{x_0, x_1, x_2, x_3, \ldots, x_{n-1}\}$. Let $y_1, y_2, \ldots, y_{n-1}$ be the inserted vertices corresponding to edges $v_0 v_i$ where $1 \leq i \leq n - 1$ and $a_1, a_2, \ldots, a_{n-2}$ be the inserted vertices on the edges $x_j x_{j+1}$ where $1 \leq j \leq n - 2$ and a_{n-1} lies in $x_{n-1} x_1$. Also the total number of edges in the $M(W_n)$ is $3(n - 1) + 1$.
Case 1: When n is even. Here $d(y_i) = n + 3(1 \leq i \leq n - 1)$
Subcase 1.1: Suppose i is odd.

Consider the set $D = \{y_i, a_{i+1}, a_{i+3}, ..., a_{n-2} \, a_{i-2}, a_{i-5}, ..., a_1\}$ where $a_j \neq N(y_i)$ and $j = 1, 3, ..., i-2, i+1, i+3, ..., n-1$ be $\gamma_{ns}(M(W_n))$ and $<V-D>$ is connected which is the minimum $\gamma_{ns}(M(W_n))$. If we place $\left(\left\lfloor\frac{n}{2}\right\rfloor\right)8+9$ pebbles on x_1, we cannot cover all the vertices of D. Hence, we require $\left(\left\lfloor\frac{n}{2}\right\rfloor\right)8+10$ pebbles to cover D. Hence, $\psi_{ns}(M(W_n)) \geq \left(\left\lfloor\frac{n}{2}\right\rfloor\right)8+10$, When n is even. Consider the distribution of $\left(\left\lfloor\frac{n}{2}\right\rfloor\right)8+10$ pebbles on the vertices of $M(W_n)$. We prove $\psi_{ns}(M(W_n)) \leq \left(\left\lfloor\frac{n}{2}\right\rfloor\right)8+10$ to cover D.

Subcase 1.2: Let the source vertex be a_k, where k is not in D.

Without loss of generality, let the source vertex be a_1. Since there will be $\left\lfloor\frac{n}{2}\right\rfloor-3$ vertices having the distance 3 and 2 vertices having the distance 2 from the source vertex then we need $\left(\left\lfloor\frac{n}{2}\right\rfloor\right)8+6 < \left(\left\lfloor\frac{n}{2}\right\rfloor\right)8+10$ Pebbles.

Subcase 1.3: Let the source vertex be x_0.

Let $p(x_0) = \left(\left\lfloor\frac{n}{2}\right\rfloor\right)8+10$. Since all the dominating vertices are at the distance of 2 from the center except y_i, we need $2 + \left\lfloor\frac{n}{2}\right\rfloor 4$ pebbles to cover the non-split dominating set. Thus, $\psi_{ns}(M(W_n)) \geq \left(\left\lfloor\frac{n}{2}\right\rfloor\right)8+10$, When n is even and i is odd.

Case 1.2: When i is even.

Consider the dominating set $D = \{y_i, a_{i+1}, a_{i+3}, ..., a_{n-1} \, a_{i-2}, a_{i-4}, ..., a_1\}$ where $a_j \neq N(y_i)$ and $j = 2, 4, ..., i-2, i+1, i+3, ..., n-1$ be the minimum dominating set and $<V-D>$ is connected. Now to prove the NSDPN of $M(W_n)$, when n is even and i is even, we can follow the same method of **Case 1, Subcase 1.1, 1.2 and 1.3.**

Case 2: When n is odd. Here $d(y_i) = n + 3(1 \leq i \leq n-1)$.

Subcase 2.1: Suppose i is odd.

Consider the set $D = \{y_i, a_{i+1}, a_{i+3}, ..., a_{n-1} \, a_{i-2}, a_{i-4}, ..., a_1\}$ where $a_j \neq N(y_i)$ and $j = 1, 3, ..., i-2, i+1, i+3, ..., n-1$ be a NSDS and $<V-D>$ is connected which is the minimum NSDS of $M(W_n)$. If we place $\left(\left\lfloor\frac{n}{2}\right\rfloor\right)8+5$ pebbles on x_1, we cannot cover all the vertices of D. Hence, we require $\left(\left\lfloor\frac{n}{2}\right\rfloor\right)8+6$ pebbles to cover D. Hence, $\psi_{ns}(M(W_n)) \geq \left(\left\lfloor\frac{n}{2}\right\rfloor\right)8+6$, When n is odd. Consider the distribution of $\left(\left\lfloor\frac{n}{2}\right\rfloor\right)8+6$ pebbles on $V(M(W_n))$. We prove $\psi_{ns}(M(W_n)) \leq \left(\left\lfloor\frac{n}{2}\right\rfloor\right)8+6$ to cover the non-split dominating set.

Subcase 2.2: Let the source vertex be a_k, where k is not in D

Without loss of generality, let the source vertex be x_1. Since there will be $\left\lfloor\frac{n}{2}\right\rfloor-2$ vertices having the distance 3, one vertex at the distance of two and one vertex is adjacent to the source vertex, then we require $\left(\left\lfloor\frac{n}{2}\right\rfloor\right)8+6$ pebbles.

Subcase 2.3: Let the source vertex be x_0.

Let us place all the pebbles on x_0. Since all the dominating vertices are at the distance of 2 from the center except y_i, we need $\left(2 + \left\lfloor\frac{n}{2}\right\rfloor\right)4$ pebbles to cover the non-split dominating set.

Subcase 2.4: Let the source vertex be any one of the vertices of y_i where i is odd.

Let us place all the pebbles on y_1. Since all the dominating vertices are at the distance of 2 from y_1 then we need $\left(1 + \left\lfloor\frac{n}{2}\right\rfloor\right)4$ pebbles to cover the non-split dominating set. Thus, $\psi_{ns}(M(W_n)) = \left(\left\lfloor\frac{n}{2}\right\rfloor\right)8+6$, When n is even and i is odd.

Case 1.2: When i is even.

Consider the dominating set $D = \{y_i, a_{i+1}, a_{i+3}, ..., a_{n-1}, a_{i-2}, a_{i-4}, ..., a_2\}$ where $a_j \neq N[y_i]$ and $j = 2, 4, ..., i-2, i+1, i+3, ..., n-1$ be the minimum dominating set and $<V - D>$ is connected. Now to prove the NSDPN of $M(W_n)$, when n is odd and i is even, we can follow the same method of **Case 2, Subcase 2.1, 2.2, 2.3 and 2.4**. Thus proved.

NSDPN for the middle graph of fan graphs
Theorem 3.4.1: Consider the middle graph of the fan. Then,

$$\psi_{ns}(M(F_n)) = \begin{cases} (\lceil \frac{n}{2} \rceil - 1)\, 8 + 6, & n \text{ is odd} \\ (\lceil \frac{n}{2} \rceil - 2)\, 8 + 6, & n \text{ is even}. \end{cases}$$

Proof: It is similar to the proof of **Theorem 3.3.1**.

References

1.   Chung, F. R. K. (1989). Pebbling in hypercubes. *SIAMJ. Disc. Math.*, 2(4), 467–472.
2.   Hurlbert, G. (1999). A survey of graph pebbling. *Congressus Numerantium*, 139, 41–64.
3.   J.A. Bondy and U.S.R. Murty, (1982), *Graph theory with applications*. USR, Elsevier Science Publishing Co., Inc. America
4.   Chartrand, G, Lesniak, L and Zhang, P, (2010), *Graphs & digraphs*, 39, CRC press.
5.   Kulli, V. R. and Janakiram, B. (2000). The non-split domination number of a graph. *Indian J. Pure Appl. Math.*, 31(4), 441–448.
6.   Crull, B., Cundiff, T., Feltman, P., Hurlbert, G. H., Pud- well, L., Szaniszlo, Z., and Tuza, Z. (2005). The cover pebbling number of graphs. *Dis. Math.*, 296(1), 15–23. https://doi.org/10.1016/j.disc.2005.03.009.

5 A model for the diagnosis and prognosis of breast cancer based on fuzzy expert system

Manisha Dubey[a] and Sanjeev Kumar

Department of Mathematics, Dr Bhimrao Ambedkar University, Khandari Campus, Agra-282002, Utar Pradesh, India

Abstract

Soft computing techniques like artificial neural networks or fuzzy inference systems have been extensively employed to mimic expert behavior in today's decision-making applications with complex concerns and knowledge involving imprecision and uncertainty. Applications for fuzzy soft computing are quickly taking off in a variety of industries, including medical diagnosis and prognosis. A number of technological researches have been published for the diagnosis of breast cancer, and some studies have also been conducted for the prognosis of breast cancer. However, the diagnosis and prognosis of breast cancer are hampered by insufficient information on the condition as well as by ambiguous and imprecise input factors. The fuzzy expert system for breast cancer diagnosis and prognosis presented in this article is able to capture the uncertainty and ambiguity that characterize breast cancer interpretation. As it is more understandable and has a high interpretability when it comes to dealing with specialists during the prognosis process, the Mamdani fuzzy inference model has been adopted in this work. The benefit of this model is that it can forecast the likelihood that females, especially those between the ages of 20 and 50, will get breast cancer. Additionally, an actual dataset is used to test the fuzzy model. This method holds promise for early cancer detection and the prediction of breast cancer risk, which could increase survival rates.

Keywords: Breast cancer, fuzzy expert system, prognosis, Mamdani fuzzy inference

Introduction

Typically, the initial stages of a cancer investigation are ambiguous and confusing. Medical professionals typically follow the patient's history and their interactions with them to make a diagnosis that is fraught with uncertainty. Therefore, it is important to have a system that can support taking decisions, whether a patient is indicated with cancer (benign or malignant) or no cancer. Here it is very much important to mention that no system can perform better than a fuzzy expert system. The main risk factors for breast cancer include age, the onset of menstruation, pregnancy or not, obesity, sedentary behavior and physical activity, smoking, and radiation exposure. A fuzzy rule-based expert system is one of the effective applications with the ability to handle uncertainty and imprecision in the diagnosis of breast cancer and its prognosis among the numerous studies reported for the detection of breast cancer.

[a]mdmanidubey@gmail.com

As far as the work done in this direction is concerned then Esogbue and Elder worked on the modeling of medical decision-making using fuzzy sets, while Zadeh wrote an article on fuzzy health, illness and disease. Kumar et al. worked on the development of continuous domain interval type-2 fuzzy sets and systems, concerning heart diseases. Mathur et al. worked on a fuzzy-based method for detecting a brain tumor in MRI images. Ali and Mutlag used fuzzy logic to detect breast cancer in its early stages. Idris and Ismail also worked on classifying breast cancer using the FUZZYDBD approach and the fuzzy-ID3 algorithm. Recently Thani and Kasbe designed a fuzzy rule-based expert system for detecting breast cancer. The age, first cycle, first pregnancy age, body mass index, smoking, and radiation exposure are the six input parameters taken into account by this work's fuzzy expert system for determining both the diagnosis and prognosis of breast cancer.

Input variables that cause the risk of breast cancer

There are some variables we have taken here that cause the risk of breast cancer in any female. We can forecast the likelihood of getting breast cancer and seek an early diagnosis by examining the effects of the variables (Age, Body Mass Index (BMI), First Menstrual Cycle (FMC), First Pregnancy Age (FPA), Smoking, and Radiation Exposure) on a female body.

Methodology

When we talk about medical sciences, we find the issues which are complex, ill-defined, or can't be easily analyzed by the medical expert so it is difficult for the expert to analyze the problem that the patient has exactly and the expert has to decide based on such information which is full of errors. When the expert analyzes the problem of the patient based on the information given by the patient, he uses his experience, common sense, and knowledge. This process is based on common sense reasoning which comes under human intelligence but nowadays medical science issues are becoming more complex so it is hard to decide on such issues as a human expert, therefore we need to design models which are based on artificial intelligence and for this, no model can perform better than a fuzzy logic-based model. For this, we have information given by the patients that is input data for the fuzzy inference system. Further, several sets of *if-then* rules can also be constructed from such knowledge. With the help of fundamental data, we aim to create a fuzzy logic-based model that will streamline decision-making and reduce the risk of cognitive diagnostic errors. For this, we take the data from the oncology department of the All India Institute of Medical Sciences, New Delhi and fuzzify the data with the help of trapezoidal membership functions. This proposed model consists of six input parameters, one output, and rules used for predicting the risk of developing breast cancer in any female. The system output is the probable decision of the risk development of breast cancer. Now based on the above information here is an opportunity to predict the risk factor by using the advancement of mathematics through fuzzy logic. Every input in the medical sciences does not result in a yes-or-no answer, or, as we might say, yes-or-no reasoning isn't particularly useful in such circumstances since the degree to which a statement is true becomes a factor. As a two-valued logic extension, fuzzy logic supports common reasoning with imprecise and ambiguous propositions involving normal language and provides a foundation for decision analysis and control actions. This system is designed to distinguish regimes in medical sciences and generates Very less, Less, Medium, High, and Very high options for a medical expert. The model is broken into four primary sections in this methodology: fuzzy rule base, fuzzy inference system, and defuzzification.

Membership function for input and output functions

Each point in the input space is mapped to a membership value between 0 and 1, which is indicated by the symbol μ, using a curve known as a membership function. As a result, the membership function must be calculated for each input and output variable.

Input functions

I.   **Age:** The input variables for the age, in years, are classified into five following fuzzy sets (VY is for the very young, Y for the young, M for the middle, O for Old age and VO for the Very Old age group):

$$\mu_{VY}(x_1) = \begin{cases} 1 & x_1 \leq 20 \\ \dfrac{30 - x_1}{10} & 20 < x_1 \leq 30 \end{cases} \tag{1}$$

$$\mu_Y(x_1) = \begin{cases} \dfrac{x_1 - 20}{10} & 20 \leq x_1 < 30 \\ 1 & 30 \leq x_1 < 40 \\ \dfrac{50 - x_1}{10} & 40 \leq x_1 \leq 50 \end{cases} \tag{2}$$

$$\mu_M(x_1) = \begin{cases} \dfrac{x_1 - 40}{10} & 40 \leq x_1 < 50 \\ 1 & 50 \leq x_1 < 60 \\ \dfrac{70 - x_1}{10} & 60 \leq x_1 \leq 70 \end{cases} \tag{3}$$

$$\mu_O(x_1) = \begin{cases} \dfrac{x_1 - 60}{10} & 60 \leq x_1 < 70 \\ 1 & 70 \leq x_1 < 80 \\ \dfrac{90 - x_1}{10} & 80 \leq x_1 \leq 90 \end{cases} \tag{4}$$

$$\mu_{VO}(x_1) = \begin{cases} \dfrac{x_1 - 80}{10} & 80 \leq x_1 < 90 \\ 1 & x_1 \geq 90 \end{cases} \tag{5}$$

II. **BMI:** The input variables for the body mass index factor are classified into three fuzzy sets (L for Low, N for Normal and H for High BMI):

$$\mu_L(x_2) = \begin{cases} 1 & x_2 \leq 15 \\ \dfrac{20 - x_2}{5} & 15 < x_2 \leq 20 \end{cases} \tag{6}$$

$$\mu_N(x_2) = \begin{cases} \dfrac{x_2 - 15}{5} & 15 \leq x_2 < 20 \\ 1 & 20 \leq x_2 < 30 \\ \dfrac{35 - x_2}{5} & 30 \leq x_2 \leq 35 \end{cases} \tag{7}$$

$$\mu_H(x_2) = \begin{cases} \dfrac{x_2 - 30}{5} & 30 \leq x_2 < 35 \\ 1 & x_2 \geq 35 \end{cases} \tag{8}$$

III. **FMC:** The input variables for the FMC factor, in years, are classified into three fuzzy sets which are as follows (E for early, N for normal and L for late FMC):

$$\mu_E(x_3) = \begin{cases} 1 & x_3 \leq 10 \\ \dfrac{10 - x_3}{2} & 10 < x_3 \leq 12 \end{cases} \tag{9}$$

$$\mu_N(x_3) = \begin{cases} \dfrac{x_3 - 10}{2} & 10 \leq x_3 < 12 \\ 1 & 12 \leq x_3 < 14 \\ \dfrac{16 - x_3}{2} & 14 \leq x_3 \leq 16 \end{cases} \tag{10}$$

$$\mu_L(x_3) = \begin{cases} \dfrac{x_3 - 14}{2} & 14 \leq x_3 < 16 \\ 1 & x_3 \geq 16 \end{cases} \tag{11}$$

IV. **FPA:** The input variables for the FPA factor, in years, are classified into three fuzzy sets which are as follows (E for early, N for normal and L for late FPA):

$$\mu_E(x_4) = \begin{cases} 1 & x_4 \leq 20 \\ \dfrac{25 - x_4}{5} & 20 < x_4 \leq 25 \end{cases} \tag{12}$$

$$\mu_N(x_4) = \begin{cases} \dfrac{x_4 - 20}{5} & 20 \leq x_4 < 25 \\ 1 & 25 \leq x_4 < 35 \\ \dfrac{40 - x_4}{5} & 35 \leq x_4 \leq 40 \end{cases} \tag{13}$$

$$\mu_L(x_4) = \begin{cases} \dfrac{x_4 - 35}{5} & 35 \leq x_4 < 40 \\ 1 & x_4 \geq 40 \end{cases} \tag{14}$$

V.   **Smoking:** The input variables for smoking factors, number of cigarettes in a day, are classified into three fuzzy sets which are as follows (*M* for medium and *H* for high smoking):

$$
\mu_L(x_s) \;=\;
\begin{cases}
1 & x_s \leq 2 \\[2mm]
\dfrac{4 - x_s}{2} & 2 < x_s \leq 4
\end{cases}
$$

$$
\mu_M(x_s) \;=\;
\begin{cases}
\dfrac{x_s - 2}{2} & 2 \leq x_s < 4 \\[2mm]
1 & 4 \leq x_s < 6 \\[2mm]
\dfrac{8 - x_s}{2} & 6 \leq x_s \leq 8
\end{cases}
\tag{15}
$$

$$
\mu_H(x_s) \;=\;
\begin{cases}
\dfrac{x_s - 6}{2} & 6 \leq x_s < 8 \\[2mm]
1 & x_s \geq 8
\end{cases}
\tag{16}
$$

VI.   **Radiation exposure:** The input variables for factor, radiation exposure, in *Gy* are classified into three fuzzy sets which are as follows (*L* for low, *M* for medium and *H* for high radiation exposure):

$$
\mu_L(x_6) \;=\;
\begin{cases}
1 & x_6 \leq 200 \\[2mm]
\dfrac{300 - x_6}{100} & 200 < x_6 \leq 300
\end{cases}
\tag{17}
$$

$$
\mu_M(x_6) \;=\;
\begin{cases}
\dfrac{x_6 - 200}{100} & 200 \leq x_6 < 300 \\[2mm]
1 & 300 \leq x_6 < 400 \\[2mm]
\dfrac{500 - x_6}{100} & 400 \leq x_6 \leq 500
\end{cases}
\tag{18}
$$

$$
\mu_H(x_6) \;=\;
\begin{cases}
\dfrac{x_6 - 400}{100} & 400 \leq x_6 < 500 \\[2mm]
1 & x_6 \geq 500
\end{cases}
\tag{19}
$$

Output function

The output of this model is denominated by strategy, which represents the expert's decision-making for the development of the risk of breast cancer prediction. The output of the linguistic values is Very High, High, Medium, Less, and Very Less.

   Case study: Considering the problem of a patient, the expert wants to know about the risk of developing breast cancer. In this case, we are using six input parameters they are *Age (x_1), BMI (x_2), FMC (x_3), FPA (x_4), Smoking (x_5),* and *Radiation Exposure (x_6).*

a)   **Input variables:** Let the input values of these variables for said patient are as follows:

$x_1 = 45$ years, $x_2 = 32$, $x_3 = 13$ years, $x_4 = 18$ years,
$x_5 = 3$ Cigarette/day, $x_6 = 280$ Gy

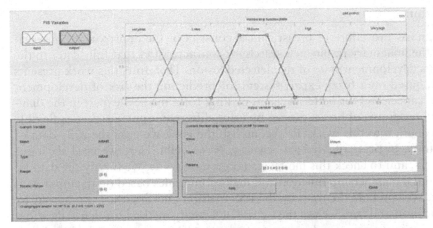

Figure 5.1 Output function based on the expert's decision-making system

1. If (Age is Very Young) and (BMI is Low) and (FMC is Early) and (FPA is Early) and (Smoking is Low) and (Radiation Exposure is Low) then (Risk is Very Less)
2. If (Age is Very Young) and (BMI is Normal) and (FMC is Early) and (FPA is Normal) and (Smoking is Low) and (Radiation Exposure is Low) then (Risk is Very Less)
3. If (Age is Young) and (BMI is Low) and (FMC is Normal) and (FPA is Early) and (Smoking is Low) and (Radiation Exposure is Low) then (Risk is Very Less)
4. If (Age is Young) and (BMI is Normal) and (FMC is Normal) and (FPA is Early) and (Smoking is Low) and (Radiation Exposure is Low) then (Risk is Very Less)
5. If (Age is Young) and (BMI is High) and (FMC is Normal) and (FPA is Early) and (Smoking is Low) and (Radiation Exposure is Medium) then (Risk is Less)
6. If (Age is Young) and (BMI is Normal) and (FMC is Normal) and (FPA is Early) and (Smoking is Low) and (Radiation Exposure is Medium) then (Risk is Less)
7. If (Age is Middle) and (BMI is Low) and (FMC is Early) and (FPA is Normal) and (Smoking is Low) and (Radiation Exposure is Low) then (Risk is Less)
8. If (Age is Very Young) and (BMI is High) and (FMC is Normal) and (FPA is Normal) and (Smoking is Low) and (Radiation Exposure is Medium) then (Risk is Medium)
9. If (Age is Very Young) and (BMI is Normal) and (FMC is Late) and (FPA is Early) and (Smoking is Medium) and (Radiation Exposure is Low) then (Risk is Medium)
10. If (Age is Young) and (BMI is High) and (FMC is Normal) and (FPA is Early) and (Smoking is Low) and (Radiation Exposure is Medium) then (Risk is Medium)
11. If (Age is Middle) and (BMI is High) and (FMC is Normal) and (FPA is Late) and (Smoking is High) and (Radiation Exposure is Medium) then (Risk is High)
12. If (Age is Old) and (BMI is High) and (FMC is Normal) and (FPA is Early) and (Smoking is Low) and (Radiation Exposure is Medium) then (Risk is Medium)
13. If (Age is Very Old) and (BMI is Normal) and (FMC is Late) and (FPA is Early) and (Smoking is Medium) and (Radiation Exposure is Medium) then (Risk is High)
14. If (Age is Middle) and (BMI is High) and (FMC is Normal) and (FPA is Early) and (Smoking is Medium) and (Radiation Exposure is Medium) then (Risk is High)
15. If (Age is Very Old) and (BMI is High) and (FMC is Late) and (FPA is Late) and (Smoking is High) and (Radiation Exposure is High) then (Risk is Very High)

Figure 5.2 Rules used for finding the output of the purposed system

b)   **Fuzzification of the crisp values of input variables:** The degree of membership of a crisp value in each fuzzy set is calculated for each linguistic variable using the membership function specified for each fuzzy set.

$$\mu_{VY}(x_1) = 0 \quad \mu_Y(x_1) = 0.5 \quad \mu_M(x_1) = 0.5 \quad \mu_O(x_1) = 0 \qquad \mu_{VO}(x_1) = 0$$
$$\mu_L(x_2) = 0 \qquad\qquad\qquad \mu_N(x_2) = 0.6 \qquad\qquad\quad \mu_H(x_2) = 0.4$$
$$\mu_E(x_3) = 0 \qquad\qquad\qquad \mu_N(x_3) = 1 \qquad\qquad\qquad \mu_L(x_3) = 0$$
$$\mu_E(x_4) = 1 \qquad\qquad\qquad \mu_N(x_4) = 0 \qquad\qquad\qquad \mu_L(x_4) = 0$$
$$\mu_L(x_5) = 0.5 \qquad\qquad\quad \mu_M(x_5) = 0.5 \qquad\qquad\quad \mu_H(x_5) = 0$$
$$\mu_L(x_6) = 0.2 \qquad\qquad\quad \mu_M(x_6) = 0.8 \qquad\qquad\quad \mu_H(x_6) = 0$$

c)   **Fire the relevant rule base for these inputs:** Fuzzy membership functions serve as the foundation for all expert systems. Applying various if-then rules, for instance, will produce the following five outputs: Very High, High, Medium, Less, Very Less.

d)   **Execute the inference system:** The "fuzzy centroid" of the composite area is calculated using the Root Sum Square approach, which scales the function at each of their individual magnitudes. Mathematically speaking, this strategy is trickier than others.

e)   **Output:** Very less risk: 0.2, less risk: 0.64, Medium risk: 0.56, High risk: 0.4, and very high risk: 0 (Figures 5.1 and 5.2)

Conclusion

The diagnosis procedure of breast cancer consists of the approximate observations of the imaginary results in a search for characteristics that fully cater to the cancerous development state of the detected lesion. Therefore, this work presents a model, based on a fuzzy expert system for predicting the risk of development of breast cancer and for better decision-making for a medical expert in the diagnosis and prognosis of breast cancer. By using different input variables, a flexible system is built into this work. To accomplish the goal, we used the daily data of the All India Institute of Medical Sciences, New Delhi. Here we used six input parameters, and to check this model we take a case study. We fuzzify the data by a trapezoidal membership function and make some rules based on the input data we took. We combine the rules by the RSS method and find the crisp output by defuzzification and here we use the Centre of Gravity method for defuzzification. Despite this imprecision and uncertainty, the model output gives the best result that an expert can have the prediction for the risk development are *Very High, High, Medium, Less, or Very Less*. The proposed model, with its probabilistic output, can be used as a support to a medical expert for better decision-making in the diagnosis of the disease. The above case study concludes that the risk of having breast cancer is medium with a degree of precision of 38.00%. The output data is then finally verified with the data of the patient available at the hospital. This has been verified and also matches the available information.

Acknowledgment

This project is funded by the UP State Government Project entitled "Mathematical Modeling of tumor growth and its treatment."

References

1. Zadeh, L. A. (1965). Fuzzy sets. *Inform. Con.*, 8, 338–353.
2. Esogbue, A. O. and Elder, R. C. (1980). Fuzzy sets and the modelling of physician decision processes, part II: Fuzzy diagnosis decision models. *Fuzzy Sets Sys.*, 3, 1–9.
3. Kosko, B. (1993). Fuzzy Thinking: The New Science of Fuzzy Logic. Hyperion: New York, 183–187.
4. Zadeh, K. S. (2000). Fuzzy health, illness and disease. *J Med. Phil.: A Forum for Bioethics and Philosophy of Medicine*, 25, 605–638.
5. Kumar, S., Sarora, M. S., and Hndoosh, R. W. (2014). The derivation of interval type-2 fuzzy sets and systems on the continuous domain: Theory and application to heart diseases. *Int. J. Sci.*, 3, 35–54.
6. Mathur, N., Mathur, S., Mathur, D., and Meena, Y. K. (2017). Detection of brain tumor in MRI image through fuzzy-based approach. *Big Data Cog. Comput.*, 27, 47–62.
7. Ali, S. K. and Mutlag, W. (2018). Early detection for breast cancer by using fuzzy logic. *J. Theoret. Appl. Inform. Technol.*, 96, 5717–5728.
8. Idris, N. F. and Ismail, M. A. (2021). Breast cancer disease classification using fuzzy-ID3 algorithm with FUZZYDBD method: Automatic fuzzy database definition. *Peer J. Comp. Sci.*, 7, e427. https://doi.org/10.7717/peerj-cs.427.
9. Thani, I. and Kasbe, T. (2022). Expert system based on fuzzy rules for diagnosing breast cancer. *Health Technol.*, 12, 473–489.

6 Error probability bounds for channel coding

K. S. Altmayer[a]

University of Arkansas, Little Rock, USA

Abstract

There has been a large interest recently in ultra reliable, low-latency wireless communication for applications in 5G and beyond which is now 6G. The usage of short block-length communication provides important applications for industry automation, machine-to-machine communication, among others including IoT (Internet of Things), and smart metering. Corresponding to this, there is renewed interest in short block-length. Recent work has strengthened earlier bounds, and they have been called the meta-converse (MC) bound and the random coding union (RCU) bound. The former places a limit on how good a code of a given rate, and block-length can be, while the latter establishes a region that is achievable. In this note we discuss the upper and lower bounds for the specific channel called binary additive white noise Gaussian (AWGN) channel model, and also a simple approximation called the normal approximation (NA) that becomes accurate for increasing block-length. The illustration is given for rate half codes with block-lengths $n=128$ and $n=256$. The work on hypotheses testing of the probability distributions and Shannon's geometrical approach is reviewed.

Keywords: AMS subject classification 94, bounds, hypothesis testing, Hausdorff space

1. Introduction

For reliable communication, binary hypothesis testing is important to the converse bounds for the error probability. The meta-converse (Polyanskiy et al., 2010) that one can use as a generic binary hypothesis test to calculate accurately the lower bound on probability in the finite block-length regime. The interest has been growing in this area of block-lengths of $n=128$ and $n=256$ also known as short packets. In this work, the main objective is to review the math modeling of the next generation wireless communication system. Shannon (1959) obtained the lower bound to the error probability of any equal power- constrained code-book using geometrical arguments including the upper bound. Polyanskiy, Poor, and Verdú (2013) applied a different approach using the lower bound for the same error probability. Their methodology is by doing binary hypothesis testing with the usage of upper and lower semi-continuity (Fan, 1953) to obtain lower and upper bounds for measurable functions in a Hausdorff spaces. These bounds are now available not only for Gaussian noise channels, i.e., additive white Gaussian noise (AWGN) with uniform distributions, but also for fading channels.

One can see details in the works of (L.E.L. and R.J.P., 2005; Polyanskiy et al., 2010) and references there. Shannon worked with the geometrical method known as sphere packing bound. Polyanskiy et al. (2010) and Polyanskiy (2013)

[a]ksaltmayer@ualr.edu

have obtained results on a tighter bound in the AWGN channel before it was obtained for capacity-achieving codes. Many previous bounds in the literature can be shown as a relaxation of the theorems obtained by the authors mentioned above. Numerical evaluations of Shannon's geometric approach are challenging. We work to complement and compare the results obtained in the present literature for the Shannon lower bound on error probability of the codes in an AWGN channel, a Gaussian and continuous applied with equal power constraints. Our main contributions are as follows:

Using the review of geometric approach of Shannon bound we provide a detailed explanation with numerical examples the Shannon's lower bound which should be the tightest bound. Then this bound is compared to the bounds, e.g., meta-converse bound obtained using probability distribution with a binary hypothesis testing. Figure 6.1 shows Shannon's geometry for a cone-packing. Shannon's lower bound is pertaining t the probability in an AWGN channel such that it moves a code-word from the n-dimensional cone which is situated at the center of the code-word and approximately uses the $1/M$ part of the outer space.

We discuss and review briefly the saddle point approximation for the two Gaussian distributions also called normal distribution in the Hausdorff spaces with the use of upper and lower semi-continuity. We compare our review of the results with the existing results. We conclude with extension to the future work in other channels and for other measurable space functions as well as non-measurable space functions which may be a product and non-product distributions in terms of Hilbert space and Banach space.

2. Minimax converse, semi-continuity, methodology

One achieves a true converse–no code can be better than the computed bound. This bound may be loose for finite lengths and asymptotically it gives

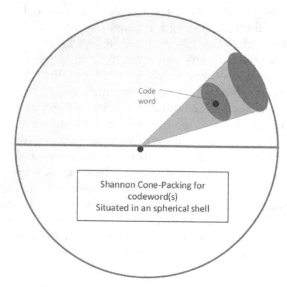

Figure 6.1 Shannon cone bound

correct results and reduces to earlier converses. The question is why have people studied more general converses for all these years? Answer–Shannon's cone bound is a good example–no specific code arrangement of points on a sphere is chosen, but the bound shows that no arrangement can be better than this bound. Many results available use the following to identify the limits of communication.

$$R_{NA} = C - \left(\sqrt{\frac{V}{n}}\right) log_2(e) Q^{-1}(P_e) + O(\frac{log_2(n)}{2n})$$ (1)

where R is the rate, C is the capacity, V is the channel dispersion and P_e is the average error probability. Normal approximation is valid for achievable and converse bounds. Polyanskiy et al. (2010) and Polyanskiy (2013) have obtained results on a tighter bound in the AWGN channel before it was obtained for capacity-achieving codes. The achievable part of the Shannon coding theorem does not address any practical capacity-approaching coding scheme. A short block-length n is considered to find the $E^*(R, n)$, the error probability. $R = k/n$ is the rate, k is the information bit. Thus, meta converse (MC) can be defined for a fixed arbitrary probability Q_Y and every (k, n, E) code must satisfy following.

$$2^k \leq \sup_{P_X}\left(\frac{1}{\beta_{1-\epsilon}(P_X P_{Y|X}, P_X Q_Y)}\right)$$ (2)

One must maximize over the probability P_X to obtain a bound that would hold for all (k, n, E). The tightness of the bound is being optimized. In other words, for a given code with a maximum-likelihood decoder, the bound given by equation (2) is tight such that the auxiliary distribution Q_Y is chosen in an optimal way. If it depends on the code itself then MC yields ML decoding probability (Vazquez-Vilar, 2019; Vazquez-Vilar et al., 2020).

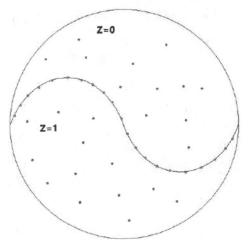

Figure 6.2 Binary hypothesis of input–output distributions

2.1. Hausdorff spaces, lower and upper semi-continuity

We consider X, and Y the two Hausdorff spaces. Further, we let $f(x, y)$ be a real-valued function on product $X \times Y$. For each x and y in X and Y respectively $f(x, y)$ is lower semi-continuous on X and $f(x, y)$ is upper semi-continuous on Y, such that the equality holds in equation (3).

The results are derived by using the convexity of the sets with probability measures on an input set and then one defines a random transformation $QX|Y$ from input to the output set. These are disjoint sets in terms of Hausdorff spaces. One would obtain a convex hull of the union of $R(P_{XY}, P_X Q_Y)$ to infer a minimum of auxiliary distribution Q_Y. The usage of probability with Neyman-Pearson lemma of hypothesis testing on two sets of probabilities with respect to Hausdorff spaces is considered. This space allows one to obtain a weaker continuity which is semi-continuous. In other words, a semi-continuous function can be defined in an Hausdorff space which will be lower semi-continuous as well as upper semi-continuous at each point of the domain of real number $\mathbf{R}$. As alluded above, P_x is the input distribution and Q_y is the output distribution. One designs Z such that input vector X has an output 0, 1 with binary value. The binary sets are with digits 0, 1. The use of KY Fan's semi-continuity provides the lower and upper bound (Figures 6.2 and 6.3). Decoding always involves a hypothesis test and looks for the property of being bounded. The upper and lower bounds involve $M = 2^k$ codewords to analyze. If f is convex on X and concave on Y, then the result of equality written as equation (3) (Fan, 1953; Polyanskiy, 2013).

2.2. Hypothesis testing, Saddle point approximation

The saddle point approximation is used for the binary AWGN channel.

2.2.1. Lemma

Lemma 2.1. *For the compact sets A and B one can achieve a saddle point approximation at a point (P^*, Q^*) in the form of following equation by using the minimax X Y theorem of (4).*

$$\min_{P_X} \sup_{Q_Y} \beta_\alpha(P_{XY}, P_X Q_Y) = \sup_{Q_Y} \min_{P_X} \beta_\alpha(P_{XY}, P_X Q_Y) \tag{3}$$

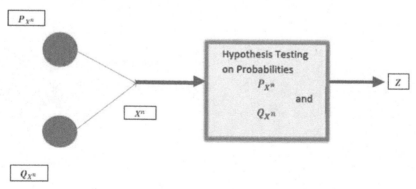

Figure 6.3 Input probability P and output probability Q

To prove it we assume that (P_X^*, Q_Y^*) is a saddle point, then we write

$$
\begin{aligned}
\min \sup \beta_\alpha(P_{XY}, P_X Q_Y) &\leq \sup \beta_\alpha(P_{XY}, P_X Q_Y) \\
&= \sup \beta_\alpha(P_{XY}, P_X Q_Y) \\
&= \min \beta_\alpha(P_{XY}, P_X Q_Y) \\
&= \sup \min \beta_\alpha(P_{XY}, P_X Q_Y)
\end{aligned}
\tag{4}
$$

Similarly, one can obtain the inequality $\sup \min \beta_\alpha(P_{XY}, P_X Q_Y) \geq \min \sup \beta_\alpha(P_{XY}, P_X Q_Y)$ This provides equality for both sides of equation (3).

2.3. Geometric approach

The claim of tightness of the bound has been studied in the papers mentioned above. It involves an expression involving a sup over Q_Y distributions, with the P_X distribution being that induced by an actual code, called P^C. We discussed this in the subsections of upper and lower semi-continuity and hypothesis testing. Now, one would need an optimization over the input distribution P_X. Shannon's 1959 paper developed a lower bound on word error probability for any code on the Gaussian channel considering a constraint that codes words lie on a sphere in n-dimensional space[1]. Shannon's argument assumes maximum likelihood decoding and is a geometric proof that lower bounds the error probability, which in general is difficult to compute exactly. Shannon's bound is a lower bound, and there is a little known about its degree of accuracy even in asymptotic cases. We investigate its tightness for short codes by pursuing the exact form of the cone bound, first for the (Shannon, 1959; Polyanskiy, 2013) single parity check code, where we know that the placement of four binary code-words on a 3-D sphere is the best. Future extensions of interest include (Fan, 1953; Vazquez-Vilar et al., 2018) extended Hamming code and the Nordstrom-Robinson code. Under the bound's suppositions, the solid angle Ω (M, n) associated with these decision zones is the area of unit n-sphere, divided by $M = 2^{nR}$. Using the known result for the area of an n-sphere we have following:

$$
\Omega(M, n) = 2^{-nR} \frac{n\pi^{n/2}}{\Gamma(n/2 + 1)} \, sterradians
\tag{5}
$$

Also Shannon's result for relating solid angle to half-cone angle θ_c is:

$$
\Omega^*(n, \theta_c) = \frac{(n-1)\pi^{(n-1)/2}}{\Gamma(n+1/2)} \int_0^{\theta_c} (sin(\alpha))^{n-2} \, d\alpha
\tag{6}
$$

Equating these two we obtain a relation between R, n and θ_c:

$$
\int_0^{\theta_c} (sin(\alpha))^{n-2} \, d\alpha = 2^{-nR} \sqrt{(\pi)} \frac{n}{n-1} \frac{\Gamma(n+1/2)}{\Gamma(n/2+1)}
\tag{7}
$$

---

[b]Upper-bounds for the best codes are also derived.

which can be solved numerically for the half cone angle as a function R and n. In the case of (Shannon, 1959; Polyanskiy, 2013) code the solutions is particularly easy. Root-finding using Newton's method seems quite practical for larger n. Following is an example for (Shannon, 1959; Polyanskiy, 2013) code.

2.3.1. An example

Example 1. An example for $M = 4$, $n = 3$, $R = 2/3$. Here we place four points on a surface of 3-sphere with vertices as tetrahedron. Ω (Fan, 1953; Polyanskiy, 2013) $= 4\pi/4 = \pi$. The cone angle is the solution.

$$\pi = \frac{2\pi}{\Gamma(2)} \int_0^{\theta_c} sin(\alpha) \, d\alpha \tag{8}$$

or

$$1/2 = (1 - cos(\theta_c)) \tag{9}$$

which leads to half cone-angle of $\pi/3$, or total cone-angle of $2\pi/3$. Continuing with the paper, Shannon notes that the probability of the received vector lands outside the cone is related to c.d.f. of the non-central t-distribution. To develop this, we observe that the deviation angle due to the noise, relative to the axis, is given by

$$\beta = cot^{-1}\frac{(\mu + n_1)}{\sqrt{\sum_2^n n_i^2}} \tag{10}$$

where n_i are i.i.d. with zero mean and unit variance, and $\mu = \sqrt{2nE_cN_0}$. We are interested in the probability that $\beta > \theta_c$:

$$P[\beta = \theta_c] = P[\frac{(\mu + n_1)}{\sqrt{\sum_2^n n_i^2}}], cot(\theta_c)]$$

$$= P[\frac{(\mu + n_1)}{\sqrt{\frac{1}{n-1}\sum_2^n n_i^2}}], cot(\theta_c)] \tag{11}$$

The quotient in the last probability expression has non-central student t-distribution, where the non-centrality parameter is $\mu = \sqrt{2nE_cN_0}$ and the degrees of freedom is $n = 1$. This probability can be easily computed using MATLAB's *nctcdf* command:

$$P[outsidecone] = nctcdf(\sqrt{n-1}cot\theta_c, n-1, \sqrt{2nE_b/N_0R}) \tag{12}$$

This becomes the lower bound for the general code with rate R and block-length n. The computational route is to first obtain θ_c and then use the above equation to obtain the lower bound.

2.4. *Normal approximation*

The normal approximation is obtained by performing an asymptotic expansion of both meta converse and random coding bound. This proves that, when suitably optimized, the bounds are asymptotically tight. This relies on the central limit theorem, and yields an approximation which is usually referred to as normal approximation (NA). It was run for the block-length $n = 24$. We compared our results to (Vazquez-Vilar et al., 2018) plot and the results seem reasonable. Please see Figures 6.4(a) and (b). The approximation of the error probability by neglecting the third term (the error term) in the dispersion is called a normal approximation (NA). The RCU and MC both can be approximated when n becomes very large by the use of NA (normal approximation).

2.4.1. *Gaussian approximation*

The NA, the normal approximation is basically Gaussian approximation. It becomes inaccurate when packets used in cell phones, other equipment, e.g., IoT (Internet of things), vehicle-to-vehicle communications, have smaller rates in comparison to the capacity of a channel. The two bounds, MC and RCU are asymptotically tight. In emerging present day applications the long code is not suitable instead short code words are better to work with latency and other type of constraints.

3. Simulation results, bounds in bi-AWGN channel

The plots of bounds are shown in Figures 6.4(a) and (b) respectively, for the bi-AWGN channel. Figure 6.4 (a) shows the plots SNR (signal to noise ratio) versus probability error for different meta converse bounds which are a lower bound, random upper bound, and Gallegar bound, also called as upper bound. Even if MC (meta converse) is not tight, it provides an intuition from the analysis of the bound itself. If a code is found such that it's closer to this code, probability will be near to the bound. Then, the auxiliary distributions Q and meta-converse are plotted. After that is the Shannon's 1959 lower bound and extended Golay code bound

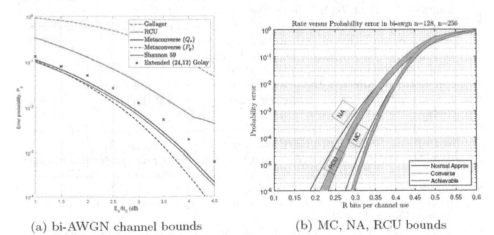

(a) bi-AWGN channel bounds   (b) MC, NA, RCU bounds

Figure 6.4 Error probability versus achievable rate

are plotted. The figure shows clearly that Shannon's bound is the tightest one. Figure 6.4(b) provides a comparison of these bounds and that the NA, the normal approximation is an approximation of MC and RCU by plotting rate versus probability error. There are two functions, one is for a fixed value of a parameter, other is for saddle point approximation using particular functions. If meta-converse is relaxed, it will not evaluate the exact error probability of the hypothesis test.

4. Conclusion

Shannon's argument assumes maximum likelihood decoding, a geometric proof, lower bounding the error probability, difficult to compute exactly. We're left with code-independent upper (RCU) and lower (MC) bounds that may not tightly bracket the performance of short codes. Polyanskiy et al. (2010) and Polyanskiy (2013) do provide the bounding that gives tighter bounds than previous bounds, and are asymptotically correct. For finite block-length "exact" performance of specific codes, we must resort to simulation, union bounds, etc. Hence in terms of 5G and now 6G, we have more work to do, an open question. There has been further development to include classical and classical-quantum sphere packing bounds (Dalai, 2017; Vazquez-Vilar, 2020).

Acknowledgment

The author Kumud Singh-Altmayer would like to thank the conference organizers at B. R. Ambedkar University and to Professor Stephen G. Wilson, University of Virginia for fruitful discussions. Further, for figure three which is similar to the Figure 6.3 in https://github.com/gdurisi/fbl-notes, permission has been obtained from the authors A. Lancho and G. Durisi. The software package was used from this website and is gratefully acknowledged.

References

1. Polyanskiy, Y., Poor, H., and Verdú, S. (2010). Channel coding rate in the finite block-length regime. *IEEE Trans. Inform. Theory*, 56, 2307–2359.
2. Shannon, C. (1959). Probability of error for optimal codes in a Gaussian channel. The Bell System Technical Journal. 38(3), 611–656. May 1959.
3. Polyanskiy, Y. (2013). Saddle point in the minimax converse for channel coding. *IEEE Trans. Inform. Theory*, 59, 2576–2595.
4. Fan, K. (1953). *Minimax Theorems*. National Bureau of Standards Report, 1–15.
5. L. E. L. and R. J. P. (2005). *Testing Statistical Hypothesis*. Berlin Heidelberg: Springer-Verlag.
6. Vazquez-Vilar, G. (2020). *Error Probability of Optimal Codes in Gaussian Channels under Average Power Constraints*. International Zürich Seminar 2020, 130–133.
7. Vazquez-Vilar, G., i Fabregas, A.G., and Verdú, S. (2019). Error probability of generalized perfect codes via meta converse. *IEEE Trans. Inform. Theory*, 65, 5706–5717.
8. Vazquez-Vilar, G., i Fabregas, A.G., Koch, T., and Lancho, A. (2018). Saddle point approximation of the error probability of binary hypothesis testing. *IEEE Int. Symp. Inform. Theory*, 2306–2310.
9. Dalai, M. (2017). *Some Remarks on Classical and Classical-Quantum Sphere Packing Bounds R`enyi vs. Kullback-Leibler*, entropy.

7 Fuzzy model for the validation and authenticity of avoid fraud at the time of the claim process

Gyan Prasad Paudel[1,a], Sanjeev Kumar[2,b] and Narayan Prasad Pahari[3,c]

[1]Graduate School of Science and Technology, Mid-West University, Surkhet, Nepal

[2]Department of Mathematics, Dr. Bhimrao Ambedkar University, Agra, India

[3]Central Department of Mathematics, Tribhuvan University, Kirtipur, Kathmandu, Nepal

Abstract

Fraud during claim settlement has become a growing concern for insurance firms in recent times. To address this issue, a fuzzy model has been developed to assist internal auditors in identifying claims that may involve fraudulent activities. The purpose of this model is to provide a structured system that helps auditors differentiate between settled claims that have indications of fraud. The fuzzy model incorporates an expert system that is designed to evaluate and validate the argument process, ensuring the credibility of the claims being made. By using fuzzy logic, which allows for imprecise and uncertain information to be handled, the model can assess the legitimacy of each claim more effectively. The model generates indicative outcomes that indicate the credibility and legitimacy of the claims being processed. These outcomes serve as a guide for auditors, enabling them to make more informed decisions during the settlement process. By utilizing this approach, auditors are better equipped to identify potentially fraudulent claims and take appropriate action. To further support the efficacy of the model, a case study is included. The case study serves as a real-life example that demonstrates the effectiveness of the fuzzy model in detecting fraudulent claims during the settlement process. The endorsement of this approach through a case study adds practical validation to the system's capabilities.

Keywords: Fuzzy logic, inference system, insurance, index of vagueness, claim validation

Introduction

Intentionally providing false information to an insurance company or agent with the aim of gaining financial benefit is known as insurance fraud. Typically, insurance fraud arises from unlawful actions committed by either the purchaser or the seller of an insurance agreement. It can also be defined as a claimant attempting to obtain some benefit or advantage to which they are not entitled, or when some due benefits are knowingly denied by an insurer. Fraudulent intentions towards an insurance provider are also common types of fraud that may occur in the course of business. Insurance fraud can occur at different points in the process and involve applicants, policyholders, third-party claimants, or

[a]gyan.math725114@gmail.com, [b]sanjeevibs@yahoo.co, [c]nppahari@gmail.com

even professionals assisting claimants. Examples of common fraudulent activities encompass exaggerating claims, providing false information on insurance applications, making claims for fictitious damages or injuries, and orchestrating staged accidents.

It is now guaranteed by individual adjustors, and this approach makes claim settlement open to arbitrary judgment. Trends show adjustors are biased against claimants and settle cases in their favor, affecting insurance monetary benefits. Moreover, both professionals and regular citizens who want to pay their deductible or see filing a claim as a chance to make a little extra money are among those who commit insurance fraud. Therefore, businesses need dependable specialist systems that can assist them in verifying or authenticating the processed statements as a result of this rising development and preventive costs on human knowledge.

A description of fuzzy pattern recognition techniques to be used in the cluster analysis of risk and classification of claims was developed by Derrig et al., while Deshmukh et al., suggested a fuzzy reasoning approach based on rules to quantify the threat of management fraud. This article illustrates the intuitive application of fuzzy sets for calculating indicators of concern, represented as red flags, on either a propositional or temporal scale. It explores the utilization of ambiguous laws to aggregate multiple red flags and outlines the creation of a specific metric for assessing the potential risk of administrative fraud. The foundation for this fuzzy logic approach is established through the utilization of established management fraud and causal models. Moreover, when it comes to claims fraud, it is an extremely uncomfortable topic for the insurance industry Brocket et al.

Unfortunately, the claims processing method can be explored by individuals and conspiratorial groups of plaintiffs and providers for their own undeserved benefit. A fuzzy-based algorithm has also been developed that enables auditors to identify fraud elements in insurance claims handled by Pathak et al. They recognized that the use of particular evaluators in the claims arbitration process provides space for independent judgment, including the use of flexibility in finalizing a lawsuit. It was discovered that claims agents were capable of resolving insurance claims on behalf of the plaintiffs merely by working with the claimant and minimizing the monetary significance of the insurance. Furthermore, Kumar et al., worked on the risk of policy cancellation by insurers applying the fuzzy inference method, where they developed a model that observes the effects of policy cancellation to avoid unwanted risk. Azar et al., have recently worked to create and validate a measure of the perception of justice using the vague principle in the context of justice theory and have shown good results in this work. The development of modern technology and cross-border communication has led to a sharp rise in fraud. Especially in sensitive and important official fields, fraud cases have increased over the past few years. A mathematical model developed by Kumar and Tiwari uses the idea of the fuzzy expert system to assess the risk of policy cancellation. To assess the likelihood of policy cancellation in the future, insurance firms can utilize a fuzzy-based expert system and fuzzy Gaussian membership function. By modifying membership roles and risk variables, this strategy aids insurance businesses in managing unpredictable occurrences and enhancing performance. Abdallah et al., sought to present a methodical and in-depth summary of these problems and difficulties that hinder the effectiveness of fraud prevention systems. A method for ranking anomalies in relation to either the majority class or two main patterns was discussed by Nian et al. It is based on a relaxation of an unsupervised spectral ranking method for anomaly problems. They demonstrated that when used with the suggested spectral ranking method for the anomaly (SRA) for fraud detection and applied to auto insurance claim data, some similarity measures, such as the Hamming distance, performed well. Fuzzy c-means (FCM) clustering stands as one of the most extensively utilized clustering techniques. To optimize cluster centroids for the detection of auto insurance fraud, Majhi introduced a hybrid clustering approach that integrates a modified whale optimization algorithm (MWOA) with FCM, employed as an under-sampling technique. In the pursuit of enhancing auto insurance fraud management and delving into the application of data mining technology in fraud detection, Yan et al., harnessed the SAGFCM-Apriori algorithm to uncover fuzzy association rules from auto insurance fraud data. These rules contribute to the identification of fraudulent claims, a critical concern given the substantial losses incurred by both insurance providers and policyholders due to fraudulent auto insurance claims. Addressing this issue, the primary goal of the research by Panda, G., et al., revolves around devising a predictive model capable of discerning the authenticity of policy claims. Furthermore, if a claim is deemed fraudulent, the study aims to pinpoint the specific parameters that warrant scrutiny to uncover such deceitful claims.

Let X be a universal set, then the collection of pairs

$$A = \{(x, \mu_A(x)): \mu_A(x): X \to [0,1], x \in X\}$$

defines a fuzzy set (Klir and Yuan) A on X.
Here, μ_A is called a membership function defined as

$$\mu_A(x) = \begin{cases} 0 & \textit{if } x \notin A \text{ and there is no ambiguity} \\ 1 & \textit{if } x \in A \text{ and there is no ambiguity} \\ (0,1) & \text{if there is ambiguity whether } x \in A \textit{ or } x \notin A \end{cases}$$

The value of $\mu_A(x)$ is the degree of element x belonging to the set A.
Let us consider three real numbers a, b, and c with $a < b < c$. The **fuzzy triangular number** X = (a, b, c) is a fuzzy number whose membership function $\mu_A(x)$ is defined as

$$\mu_X(x) = \begin{cases} \dfrac{x-a}{b-a} & \textit{if } x \in [a,b] \\ \dfrac{c-x}{c-b} & \textit{if } x \in [b,c] \\ 0 \ \textit{if } a > x, & \textit{or } x > c \end{cases}$$

We note that, in the above definition, we have $\mu_A(x) = 1$, while b need not be the mid-point of a and c.
A fuzzy number $A = (a, b, c, d)$ is said to be a **trapezoidal fuzzy number** with the membership function defined by

$$\mu_A(x) = \begin{cases} 0 & for\ x \le a \\ \frac{x-a}{b-a} & for\ a \le x \le b \\ 1 & for\ b \le x \le c \\ \frac{d-x}{d-c} & for\ c \le x \le d \\ 0\ for\ otherewise \end{cases}$$

Fuzzification involves assigning the numerical input of a system to fuzzy sets with a certain level of membership. This level of membership can take on any value between 0 and 1. A value with a membership degree of 0 indicates that it does not belong to the given fuzzy set, while a membership degree of 1 signifies complete inclusion within the fuzzy set. So when the degree of membership of an element is 0, it is not considered part of the fuzzy set; conversely, when the degree of membership is 1, the element is entirely encompassed by the fuzzy set. Any number between 0 and 1 indicates how uncertainly the value belongs within the set. Due to the fact that these fuzzy sets are often characterized by words, we may reason with them in a linguistically reasonable way by associating the system input with fuzzy sets.

Defuzzification refers to the procedure of converting fuzzy sets and their corresponding membership degrees into a well-defined outcome within crisp logic. This process entails converting a fuzzy set into a crisp set, resulting in a measurable and distinct outcome.

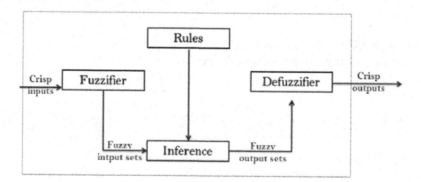

2. Key features

The proposed method is capable of testing claims of any size. The professional fuzzy logic insurance claim fraud detection scheme is based on the "imprecision index" idea. Fuzzy logic seems to be a very helpful method of coping with uncertainty, approximation, inaccuracy, misunderstanding, inaccuracy, or causes of inaccuracy that are not statistics in the human reasoning and decision-making process. We may calculate the relevance of a collection of information to different criteria of unclear significance by introducing fuzzy logic. If Δ_{ij} is the variable $0–1$ ($\Delta_{ij} = 1$, if there is any discrepancy/change between the information supplied by the plaintiff and that obtained by the auditor, otherwise zero) and W_{ij} is the weighting or impact factor for the ith section for the jth information, then the inaccuracy index for the ith section of the information is provided by Ross (1997).

$$X_i = \frac{\left(\Sigma_{i=1}^I \Sigma_{j=1}^J W_{ij}\Delta_{ij}\right)}{I} \tag{1}$$

It must be remembered that the details critical in the decision to solve the argument, as well as all weightings for a collection of i-th data $\Sigma_{j=1}^J W_{ij}$ to the unit added, are given a greater weighting/effect factor. Likewise, it is possible to determine the values of the other entries. A triangular association function is used here for the "degree of incompleteness" input variable, while trapezoidal association functions are used for the "inaccuracy index", "rating level" and "claimant's credit report" input variables. The claim must be verified and validated when handling the claim in order to reduce the relation time of settlements (Figure 7.1).

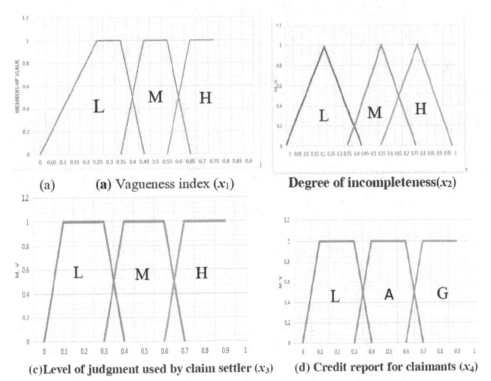

(a)   (a) Vagueness index (x_1)   Degree of incompleteness(x_2)

(c)Level of judgment used by claim settler (x_3)   (d) Credit report for claimants (x_4)

Figure 7.1 Membership functions of inputs. (a) Vagueness index (x_1). (b) Degree of incompleteness (x_2). (c) Level of judgment used by claim settler (x_3). (d) Credit report for claimants (x_4).

3. Algorithm

i.   **Input:** Clear value of settlement of and other collected data.

ii. **Compare with the limit:** Whether the settlement benefit of the lawsuit, as calculated by the auditors, is less than a certain predetermined sum(say alpha = Rs. 2,000,00), so the settled claim can be deemed authentic and proceed to stage 8.

iii. **Incoming rating:** Rating the inaccuracy index x_1, the level of the incompleteness of the claims x_2, the rating level used by claim settlers x_3, and the claimants' credit report x_4.

iv. **Crisp input values are fuzzified:** To determine the level of involvement of each fuzzy set with a crisp value, use the membership functions supplied to each fuzzy set for each linguistic variable. The membership equation for computation is:

$$f(x, a, b, c) = \begin{cases} \frac{x-a}{b-a} & if\ a \leq x \leq b \\ 1 & if\ b \leq x \leq c \\ \frac{d-x}{d-c} & if\ c \leq x \leq d \end{cases} \tag{2}$$

$$\mu_L(x_1) = \begin{cases} \frac{x_1-0}{0.25} & if\ 0 \leq x_1 \leq 0.25 \\ 1 & if\ 0.25 \leq x_1 \leq 0.35 \\ \frac{0.45-x_1}{0.10} & if\ 0.35 \leq x_1 \leq 0.45 \end{cases} \tag{3}$$

$$\mu_M(x_1) = \begin{cases} \frac{x_1-0.35}{0.10} & if\ 0.35 \leq x_1 \leq 0.45 \\ 1 & if\ 0.45 \leq x_1 \leq 0.55 \\ \frac{0.65-x_1}{0.10} & if\ 0.55 \leq x_1 \leq 0.65 \end{cases} \tag{4}$$

$$\mu_H(x_1) = \begin{cases} \frac{x_1-0.55}{0.25} & if\ 0.55 \leq x_1 \leq 0.65 \\ 1 & if\ x_1 \geq 0.65 \end{cases} \tag{5}$$

$$\mu_L(x_2) = \begin{cases} \frac{x_2}{0.23} & if\ x_2 \leq 0.23 \\ \frac{0.45-x_2}{0.22} & if\ 0.23 \leq x_2 \end{cases} \tag{6}$$

$$\mu_M(x_2) = \begin{cases} \frac{x_2-0.35}{0.20} & if\ x_2 \leq 0.55 \\ \frac{0.74-x_2}{0.20} & if\ 0.55 \leq x_2 \end{cases} \tag{7}$$

$$\mu_H(x_2) = \begin{cases} \frac{x_2-0.55}{0.21} & if\ x_2 \leq 0.76 \\ \frac{1-x_2}{0.24} & if\ 0.76 \leq x_2 \end{cases} \tag{8}$$

$$\mu_L(x_3) = \begin{cases} \frac{x_3-0}{0.15} & if\ 0 \leq x_3 \leq 0.15 \\ 1 & if\ 0.15 \leq x_3 \leq 0.30 \\ \frac{0.45}{0.15} & if\ 0.30 \leq x_3 \leq 0.45 \end{cases} \tag{9}$$

$$\mu_M(x_3) = \begin{cases} \frac{x_3 - 0.30}{0.15} & if \quad 0.30 \le x_3 \le 0.45 \\ 1 & if \quad 0.45 \le x_3 \le 0.60 \\ \frac{0.75 - x_3}{0.15} & if \quad 0.60 \le x_3 \le 0.75 \end{cases} \tag{10}$$

$$\mu_H(x_3) = \begin{cases} \frac{x_3 - 0.60}{0.15} & if \quad 0.60 \le x_3 \le 0.75 \\ 1 & if \qquad x_3 \ge 0.75 \end{cases} \tag{11}$$

$$\mu_L(x_4) = \begin{cases} \frac{x_4 - 0}{0.10} & if \quad 0 \le x_4 \le 0.10 \\ 1 & if \quad 0.10 \le x_4 \le 0.30 \\ \frac{0.4 - x_4}{0.10} & if \quad 0.30 \le x_4 \le 0.40 \end{cases} \tag{12}$$

$$\mu_A(x_4) = \begin{cases} \frac{x_4 - 0.30}{0.10} & if \quad 0.30 \le x_4 \le 0.40 \\ 1 & if \quad 0.40 \le x_4 \le 0.60 \\ \frac{0.70 - x_4}{0.10} & if \quad 0.60 \le x_4 \le 0.70 \end{cases} \tag{13}$$

$$\mu_G(x_4) = \begin{cases} \frac{x_4 - 0.60}{0.10} & if \quad 0.60 \le x_4 \le 0.70 \\ 1 & if \qquad x_4 \ge 0.70 \end{cases} \tag{14}$$

Here, (a, b, c, d) = vertices of the functions of the trapezoidal membership, (a, b, c) = vertices of the functions of the triangular membership, L, M, H, A, and G represent the fuzzy collection for low, medium, high, average, and good, respectively.

v.   **Trigger the rule bases corresponding to these inputs**
"If-then" rules that are based on fuzzy logic are used by all expert systems. The "if" portion is linked to this as a precedent as well as an assumption, although the "then" portion is known to it as the end result or conclusion. Even though three parameters possess three fuzzy sets (low, medium, and high) and another fuzzy set (low, average, and good), it is, therefore, appropriate to fire 81 (3×3×3×3) fuzzy decisions. There are two outputs: a suspect's authentic factor-AS (makes reference to damage claims that have been honestly resolved with little space besides vagueness or suspicion) and "cases settled with fraud element - SF" (corresponds to the claims resolved but containing questionable components that would better be substantially scrutinized).

vi.   **Perform the inference engine**
The inference mechanism may also bind the fuzzy rule of a fuzzy inference method to derive a numeric value for both the intermediate and the output variable as early as all crunchy input data has already fuzzified into their respective numeric variables. Aggregation and composition are the two key steps in the process of inference. The aggregation step includes the calculation of the principles of the component, including its laws, if (antecedent) has been the method of computing the rate of the component of the standards, then (conclusion). Through aggregation, a level of reality dependent on the membership function of a consequent linguistic variable is assigned to each form in the case of a part of a law. From here, to clip the amount of truth

from all of the if portions, the result (PROD) of the level of truth of the conditions is determined. This is referred to as the component's degree of fact. The next step of the inference process is to decide the degree of truth of the output linguistic variable for each linguistic expression. In general, for each linguistic sign of the frequency linguistic word, to determine the degrees of certainty, the highest (max) or complete (sum) degrees of the truth of the method are determined with the same linguistic variables in the sections below.

vii.   **Defuzzification**

A recent stage in the fuzzy inference method has been the defuzzification of a linguistic description of the output linguistic parameters into numerical values. The much more popular defuzzification form, center-of-maximum (CoM), can be used for production defuzzification, although one could go into (center-of-area (CoA). At first, CoM assesses the much greater meaning for an output linguistic word as well as the linguistic component and then calculates the crisp value as the proper method for both principles and related degrees of association given.

viii.   **The output of decisions of the expert system**

The development forms, in this case, are the true settlement of the lawsuit and charges settled with a defendant or part of the scam. Each controller's basic characteristics are based on the model and the computation of efficiency. Theoretically, however, in all the fuzzy logic-dependent expert methods, we evaluate the tacit and explicit relationships within the system by replicating human reasoning and finally establishing the optimum fuzzy management rules and the database (Figure 7.2).

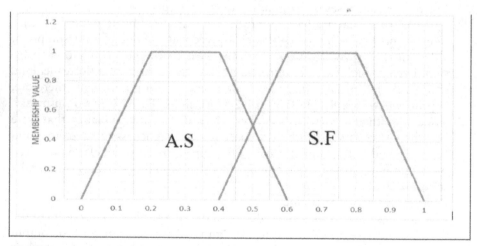

Figure 7.2 Membership function of output (Authentic settlement/ fraud settlement)

4. Case study

For instance, we assume that four inputs are used by the insurance provider, namely the vagueness indicator x_1 the extent of incomplete claims x_2, the amount of judgment used among claim invaders x_3, and the credit report of a claimant x_4. Such measures were also indicative of the provenance of an insurance payout settlement. The degree of vagueness/doubt with data as an indication of an inaccurate insurance settlement reflects these inputs and the extent of control used by the claimants to assess the settlement.

i.   **Input: Value of the claim** = Rs. 80,000.00
ii. **Compare against a threshold**: Although the value is higher than the present value (Rs. 25,000.00). The system of experts may review the validity of the settled claim.
iii. **Test the validity of claim settlement**: The value of the statements' inputs must be measured, $x_1 = 0.38, x_2 = 0.70, x_3 = 0.65, x_4 = 0.62$ (say)
iv. **Fuzzification of inputs crisp values**: Though using the membership characteristics defined for language variable for each fuzzy package. The degree of membership in the fuzzy set of crisp values is defined as follows:

$$\mu_L(x_1) = \frac{0.45 - x_1}{0.10} = 0.70 \tag{15}$$

$$\mu_M(x_1) = \frac{x_1 - 0.35}{0.10} = 0.30 \tag{16}$$

$$\mu_M(x_1) = 0 \tag{17}$$

$$\mu_L(x_2) = \max\left\{0, \frac{0.45 - x_2}{0.22}\right\} = 0 \tag{18}$$

$$\mu_M(x_2) = \max\left\{0, \frac{0.75 - x_2}{0.20}\right\} = 0.25 \tag{19}$$

$$\mu_H(x_2) = \max\left\{0, \frac{x_2 - 0.55}{0.21}\right\} = 0.667 \tag{20}$$

$$\mu_L(x_3) = 0 \tag{21}$$

$$\mu_M(x_3) = \frac{0.75 - x_3}{0.15} = 0.67 \tag{22}$$

$$\mu_H(x_3) = \frac{x_3 - 0.60}{0.15} = 0.33 \tag{23}$$

$$\mu_L(x_4) = 0 \tag{24}$$

$$\mu_M(x_4) = \frac{0.70 - x_4}{0.10} = 0.80 \tag{25}$$

$$\mu_H(x_4) = \frac{x_4 - 0.60}{0.10} = 0.20 \tag{26}$$

v.   **Trigger the rule bases corresponding to these inputs**: The following rules, based on the importance of the fuzzy membership function values, refer to the example under consideration.

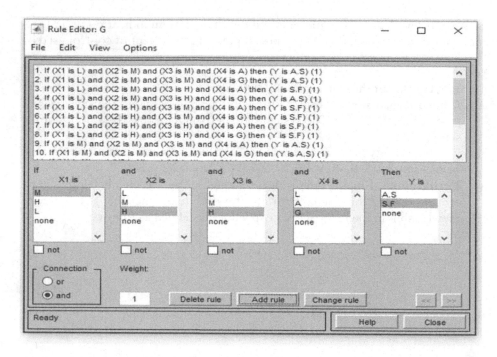

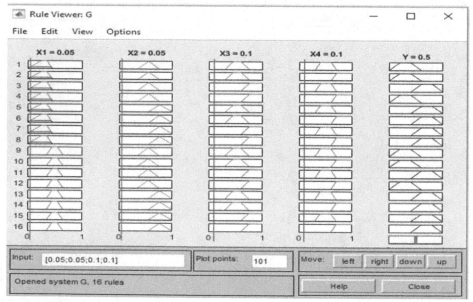

vi. **Execute the engine of inference:** We see the "root sum squares" (RSS) types to combine the results of all relevant laws and scale the functions to their respective strengths (range [0,1]) of the output membership function from possible rules (R 1- 81) are :

$$A.S = \sqrt{(0.25)^2 + (0.20)^2 + (0.20)^2 + (0.25)^2 + (0.20)^2 + (0.20)^2}$$

$$= \sqrt{0.2850}$$

$$= 0.53$$

$$S.F = \sqrt{(0.25)^2 + (0.667)^2 + (0.20)^2 + (0.33)^2 + (0.20)^2 + (0.25)^2 + (0.30)^2 + (0.20)^2 + (0.30)^2 + (0.20)^2}$$

$$= \sqrt{1.0187}$$

$$= 1.0(approx.)$$

vii. **Defuzzifucation:** By incorporating the outcomes of an evaluation procedure, we just use the "weighted average approach" to achieve the defuzzification of such information into the crisp output. By measuring almost every membership function in the outcome by the corresponding maximum membership value, the weighted average technique was created and the result is called the crisp outcome. 0.48 is the crisp performance. The crisp production belongs more to the SF collection than to the true settlement set (as obvious with its basis functions). The judgment in this case, then, is SF.

viii. **Through the expert system assessment:** The process means that misconduct has been fixed in the case of the proceedings under examination. In the above example, it is in between A. S and S. F. but towards S. F. So we will give only the threshold amount which is Rs. 2500.00

5. Result and discussion

The study provides a fuzzy model to assist internal auditors in distinguishing between all resolved claims including an element of fraud. The model is intended to validate and validate the argument process, and it offers indicative results for the credibility and validity of the statement in the process. A case study is also used to support the strategy. The analysis yielded a sharp result of 0.48, which belongs more to the SF collection than to the genuine settlement set. In this situation, the decision is SF. The two outputs of the model are a suspect's authentic factor AS and "cases settled with fraud element SF." The importance of fraud prevention in the insurance sector is highlighted in the paper's discussion, along with how the fuzzy model might help auditors spot false claims. The report also addresses the model's shortcomings and makes suggestions for future studies to enhance its efficacy.

6. Conclusion

To measure the indicative outcome of the validation/authenticity of the argument using four variables, the whole article gives a fuzzy rule-dependent method. These four main factors that play a vital role in this area have restricted our work. It should be noted that adding more variables will also improve the performance of fuzzy rules to the level that tuning its rule base utilizing real scenario information is absolutely required in order to enhance the efficiency of the model. We propose that from a few of the tried combinations, the use of the neural network, genetic algorithm, and MATLAB would then yield a maximum layer chosen to represent all mixture points. Although this method cannot be used to detect fraud during the processing of insurance claims, there are some pieces of research done in that field, the formation of a neural network (NN)-dependent method to detect organizational fraud by using financial information published has been documented.

References

1. Abdallah, A., Maarof, M. A., and Zainal, A. (2016). Fraud detection system: A survey. *J. Netw. Comp. Appl.*, 68, 90–113.
2. Azar, A. and Darvishi, Z. A. (2011). Development and validation of a measure of justice perception in the frame of fairness theory-fuzzy approach. *J. Exp. Sys. Appl.: Int. J.*, 38(6), 7364–7372.
3. Brockett, P. L., Xia, X., and Derrig, R. A. (1998). Using Kohonen's self-organizing feature map to uncover automobile bodily injury claims fraud. *J. Risk Ins.*, 245–274.
4. Derrig, R. and Ostaszewski, K. (1995). Fuzzy techniques of pattern recognition in risk and claim classification. *J. Risk Ins.*, 62(3), 447–482.
5. Deshmukh, A. and Talluru, L. (1998). A rule-based fuzzy reasoning system for assessing the risk of management fraud. *Intel. Sys. Account. Fin. Manag.*, 7(4), 223–241.
6. Klir, G. and Yuan, B. (1995). Fuzzy sets and fuzzy logic. New Jersey: Prentice Hall. 4, 1–12.
7. Kumar, S. and Pathak, P. (2009). Premium allocation-fuzzy approach in insurance business. Proceeding of the 3rd National Conference, INDIA Com., 703–706.
8. Kumar, S. and Tiwari, N. (2015). Mathematical model for the risk of cancellation of life insurance policies. *SRM Int. J. Engg. Sci.*, 19.
9. Majhi, S. K. (2021). Fuzzy clustering algorithm based on modified whale optimization algorithm for automobile insurance fraud detection. *Evol. Intel.*, 14(1), 35–46.
10. Nian, K., Zhang H., Tayal, A., Coleman, T., and Li, Y. (2016). Auto insurance fraud detection using unsupervised spectral ranking for the anomaly. *J. Fin. Data Sci.*, 2(1), 58–75.
11. Panda, G., Dhal, S. K., Satpathy, R., and Pani, S. K. (2022). ANFIS for fraud automobile insurance detection system. *Adv. Data Sci. Manag.*, 519–530.
12. Pathak, J., Vidhyarthi, N., and Summers, S. L. (2005). A fuzzy-based algorithm for auditors to detect elements of fraud in settled insurance claims. *Manag. Audit. J.*, 20(6), 632–644.
13. Ross, T. J. (1997). Fuzzy Logic with Engineering Application. Singapore: McGraw-Hill.
14. Yan, C., Liu, J., Liu, W., and Liu, X. (2021). Research on automobile insurance fraud identification based on fuzzy association rules. *J. Intel. Fuzzy Sys.*, 1–14.

8 The combined effects of thermal and mass stratification on unsteady flow past an exponentially accelerated vertical plate with variable temperature

Rupam Shankar Nath[a], Rudra Kanta Deka and Himangshu Kumar

Department of Mathematics, Gauhati University, Assam, India

Abstract

This research paper investigates the impact of thermal and mass stratification on unsteady flow through a vertical plate that accelerates exponentially with variable temperature. The method of the Laplace transform is utilized to solve the governing equations for unit Prandtl and Schmidt number. The results showed that both thermal and mass stratification have considerable influence on the flow's velocity, temperature, and concentration fields. We obtained some significant results when we compared the stratification case with the no stratification case. The results of this study can be applied to industrial systems that have unsteady flows with stratified fluids to enhance their design and performance.

Keywords: Thermal stratification, mass stratification, exponentially accelerated, unsteady flow

Introduction

The study of stratification is essential to the fields of fluid dynamics and engineering, because a deeper comprehension of this phenomenon may lead to the creation of machinery that is more effective and efficient. This study is essential for understanding the connection between temperature, velocity, and density, as well as how these three components interact as fluid flows past a vertical plate that accelerated exponentially. The research article will examine the fluid's behavior under various temperature and mass stratification circumstances to determine which one produces the best outcome. The consequences of the findings and how they may be applied in practical applications will also be discussed in the research paper.

Cheng (2008) and Cheng (2009) looked into the impacts of thermal and mass stratifications over a vertical wavy truncated cone and a wavy surface, respectively. Moreover, Deka and Paul (2013) and Paul and Deka (2017) have studied how both effects affect infinite vertical cylinders. The impact of thermal and mass stratification on unstable flow through an infinite plate has never been studied before, and this study is the first to do so. Magyari et al. (2006), Bhattacharya and Deka (2011) and Shapiro and Fedorovich (2004) investigated unsteady flows in a Stably Stratified Fluid, focusing on infinite plates with thermal stratification only.

---

[a]rupamnath23@gmail.com

Furthermore, buoyancy-driven flows in a stratified fluid were examined by Park and Hyun (1998) and Park (2001). Das et al. (1994) came up with an analytical solution to describe how fluid would flow past an infinite vertical plate that had been affected chemically. Vemula et al. (2006), Chamkha et al. (2016) and Ramakrishna et al. (2022) are investigated with distinct effects over fluid passing through a vertical plate that accelerates exponentially. Kalita et al., (2023) and Nath and Deka (2023) conducted study on the impacts of thermal stratification and chemical reactions on the unsteady flow through different types of vertical plates, respectively.

Mathematical analysis

Consider the phenomenon of an unstable flow of a viscous in-compressible stratified fluid past a vertical plate that accelerates exponentially with fluctuating temperature. In this problem, the x' axis is considered vertically upward along the plate, whereas the y' axis is taken normally to the plate. The temperature and concentration of the fluid and plate are equal at time $t' = 0$, with T_∞' and C_∞', respectively. At time $t' > 0$ the temperature of the plate exhibits a linear relationship with respect to the variable of time, t and concentration increase to C_ω'. All flow variables are unrelated of x' and solely depend on y' and t' due to the infinite of the plate. Hence, we get a 1-D flow field with just one non-zero vertical velocity component u'. The unsteady flow is therefore determined by the resulting equations, using the typical Boussinesq approximation.

$$\frac{\partial u'}{\partial t'} = g\beta(T' - T_\infty') + g\beta^*(C' - C_\infty') + v\frac{\partial^2 u'}{\partial y'^2} \tag{1}$$

$$\frac{\partial T'}{\partial t'} = \alpha\frac{\partial^2 T'}{\partial y'^2} - \gamma u' \tag{2}$$

$$\frac{\partial C'}{\partial t'} = D\frac{\partial^2 C'}{\partial y'^2} - \xi u' \tag{3}$$

With the given boundary and initial conditions are as follows:

$$
\begin{aligned}
u' &= 0 & T' &= T_\infty' & C' &= C_\infty' & \forall y', t' \le 0 \\
u' &= e^{a't'} & T' &= T_\infty' + (T_\omega' - T_\infty')At' & C' &= C_\omega' & \text{at } y' = 0, t' > 0 \\
u' &= 0 & T' &\to T_\infty' & C' &\to C_\infty' & \forall y' \to \infty, t' > 0
\end{aligned}
$$

The "thermal stratification" and "mass stratification" are termed as $\gamma = \frac{dT_\infty'}{dx'} + \frac{g}{c_p}$ and $\xi = \frac{dC_\infty'}{dx'}$, respectively. The term "thermal stratification" refers to the amalgamation of vertical temperature advection $\frac{dT_\infty'}{dx'}$, where the temperature of the surrounding fluid is height-dependent, and work of compression $\frac{g}{c_p}$, the rate at which particles in a fluid do reversible work due to compression.

Non-dimensional quantities are as follows:

$$U = \frac{u'}{u_0}, \quad t = \frac{t'u_0^2}{v}, \quad y = \frac{y'u_0}{v}, \quad \theta = \frac{T' - T'_\infty}{T'_\omega - T'_\infty}, \quad C = \frac{C' - C'_\infty}{C'_\omega - C'_\infty}, \quad Gr = \frac{g\beta(T' - T'_\infty)}{u_0^3}$$

$$Gc = \frac{g\beta^*(C' - C'_\infty)}{u_0^3}, \quad Pr = \frac{v}{\alpha}, \quad Sc = \frac{v}{D}, \quad a = \frac{va'}{u_0^2}, \quad S = \frac{\gamma v}{u_0(T'_\omega - T'_\infty)}, \quad F = \frac{\xi v}{u_0(C'_\omega - C'_\infty)}$$

$A = \frac{u_0^2}{v}$ is the constant.

The non-dimensional forms of the equations (1–3) are given by

$$\frac{\partial U}{\partial t} = Gr\,\theta + Gc\,C + \frac{\partial^2 U}{\partial y^2}$$

$$\frac{\partial \theta}{\partial t} = \frac{1}{Pr}\frac{\partial^2 \theta}{\partial y^2} - SU$$

$$\frac{\partial C}{\partial t} = \frac{1}{Sc}\frac{\partial^2 C'}{\partial y^2} - FU$$

Boundary and initial conditions in absence of dimensional form are:

$$U = 0 \qquad \theta = 0 \qquad C = 0 \qquad \forall y, t \le 0$$
$$U = e^{at} \qquad \theta = t \qquad C = 1 \qquad at \quad y = 0, t > 0$$
$$U = 0 \qquad \theta \to 0 \qquad C \to 0 \qquad at \quad y \to \infty, t > 0$$

Method of solution

The solutions for velocity, temperature, and concentration profiles can be determined with the help of Abramowitz ad Stegun (1972) and Hetnarski (1975) for $Pr = 1, Sc = 1$. Hence, we get

$$U = \frac{1}{2}[f_3(iB) + f_3(-iB)] - \frac{Gr}{2iB}[f_2(iB) - f_2(-iB)] - \frac{Gc}{2iB}[f_1(iB) - f_1(-iB)]$$

$$\theta = \frac{tFGc}{B^2}\left[(1 + 2\eta^2)erfc(\eta) - \frac{2\eta}{\sqrt{\pi}}e^{-\eta^2}\right] - \frac{SGc}{B^2}erfc(\eta) + \frac{S}{2iB}[f_3(iB) - f_3(-iB)]$$

$$+ \frac{SGr}{2B^2}[f_2(iB) + f_2(-iB)] + \frac{SGc}{2B^2}[f_1(iB) + f_1(-iB)]$$

$$C = \frac{SGr}{B^2}erfc(\eta) - \frac{tFGr}{B^2}\left[(1 + 2\eta^2)erfc(\eta) - \frac{2\eta}{\sqrt{\pi}}e^{-\eta^2}\right] + \frac{F}{2iB}[f_3(iB) - f_3(-iB)]$$

$$+ \frac{FGr}{2B^2}[f_2(iB) + f_2(-iB)] + \frac{FGc}{2B^2}[f_1(iB) + f_1(-iB)]$$

Where, $\eta = \frac{y}{2\sqrt{t}}$ $\quad B = \sqrt{SGr + FGc}$.

Furthermore, the inverse Laplace transformations of f_i's are provided by

$$f_1(ip) = L^{-1}\left\{\frac{e^{-y\sqrt{s+ip}}}{s}\right\} \quad f_2(ip) = L^{-1}\left\{\frac{e^{-y\sqrt{s+ip}}}{s^2}\right\} \quad f_3(ip) = L^{-1}\left\{\frac{e^{-y\sqrt{s+ip}}}{s-a}\right\}.$$

We separate the complex arguments of the error function contained in the previous expressions into real and imaginary parts using the formulas provided by Abramowitz and Stegun (1972).

Special case (S = 0, F = 0)
We derived solutions for the special case of no thermal and mass stratification ($S = 0$, $F = 0$) we want to compare the results of the fluid with thermal and mass stratification to the case with no stratification. Hence, the corresponding solutions for the special case are given by:

$$U^* = \frac{e^{at}}{2}\left[e^{-2\eta\sqrt{at}}erfc(\eta - \sqrt{at}) + e^{2\eta\sqrt{at}}erfc(\eta + \sqrt{at})\right] + 2\eta t Gc\left[\frac{e^{-\eta^2}}{\sqrt{\pi}} - \eta erfc(\eta)\right]$$
$$+ \frac{\eta t^2 Gr}{3}\left[\frac{4}{\sqrt{\pi}}(1 + 2\eta^2)e^{-\eta^2} - \eta(6 + 4\eta^2)erfc(\eta)\right]$$

$$\theta^* = t\left[(1 + 2\eta^2)erfc(\eta) - \frac{2\eta}{\sqrt{\pi}}e^{-\eta^2}\right]$$

$$C^* = erfc(\eta).$$

Skin-friction
The non-dimensional skin-friction, which is determined as shear stress on the surface, is represented by as follows

$$\tau = \frac{\cos Bt}{\sqrt{\pi t}} + e^{at}(P_1 r_3 - Q_1 r_4) + \frac{Gr}{B}\left[\sqrt{\frac{t}{\pi}}\sin Bt - t\sqrt{\frac{B}{2}}\cdot(r_1 + r_2) + \frac{(r_1 - r_2)}{2\sqrt{2B}}\right] + \frac{Gc}{B}\left[\frac{\sin Bt}{\sqrt{\pi t}} - \sqrt{\frac{B}{2}}(r_1 + r_2)\right].$$

The skin-friction for the special case is given by

$$\tau^* = e^{at}\sqrt{a}\ erf(\sqrt{at}) + \frac{3 - 2t^2 Gr}{3\sqrt{\pi t}} - Gc\sqrt{\frac{t}{\pi}}.$$

Nusselt number
The non-dimensional Nusselt number, which is determined as the rate of heat transfer, is represented by as follows

$$Nu = \frac{2FGc}{B^2}\sqrt{\frac{t}{\pi}} - \frac{SGc}{B^2}\frac{1}{\sqrt{\pi t}} - \frac{S}{B}\left[\frac{\sin Bt}{\sqrt{\pi t}} - e^{at}(P_1 r_4 + Q_1 r_3)\right] + \frac{SGc}{B^2}\left[\frac{\cos Bt}{\sqrt{\pi t}} + \sqrt{\frac{B}{2}}(r_1 - r_2)\right]$$
$$+ \frac{SGr}{B^2}\left[\sqrt{\frac{t}{\pi}}\cos Bt + t\sqrt{\frac{B}{2}}(r_1 - r_2) + \frac{(r_1 + r_2)}{2\sqrt{2B}}\right].$$

The Nusselt number for the special case is given by $Nu^* = 2\sqrt{\frac{t}{\pi}}$.

Sherwood number
The non-dimensional Sherwood number, which is determined as the rate of mass transfer, is represented by as follows

$$Sh = \frac{SGr}{B^2}\frac{1}{\sqrt{\pi t}} - \frac{2FGr}{B^2}\sqrt{\frac{t}{\pi}} - \frac{F}{B}\left[\frac{\sin Bt}{\sqrt{\pi t}} - e^{at}(P_1 r_4 + Q_1 r_3)\right] + \frac{FGc}{B^2}\left[\frac{\cos Bt}{\sqrt{\pi t}} + \sqrt{\frac{B}{2}}\,(r_1 - r_2)\right]$$

$$+ \frac{FGr}{B^2}\left[\sqrt{\frac{t}{\pi}}\cos Bt + t\sqrt{\frac{B}{2}}\,(r_1 - r_2) + \frac{(r_1 + r_2)}{2\sqrt{2B}}\right].$$

The Sherwood number for the special case is given by $Sh^* = \dfrac{1}{\sqrt{\pi t}}$.

Where, $erf(\sqrt{iBt}) = r_1 + ir_2$, $erf(\sqrt{(a + iB)t}) = r_3 + ir_4$ and $\sqrt{a + iB} = P_1 + iQ_1$.

Results and discussion

The outcomes derived from the numerical computations were displayed in diagrams to explain the physical significance of the situation. In the graphs, the relationships between the physical components Gr, Gc, S, F and time t were shown, together with variations in concentration, temperature, velocity, rate of momentum transmission, transfer of heat and mass transfer rate. We represented them in graphs in Figures 8.1–8.14.

The impacts of thermal (S) and mass stratification (F) on the velocity are represented in Figure 8.1. It is shown that the velocity is decreased by both thermal and mass stratification. We found that fluid velocity decreases for mass stratification $F > 0$, as it does for thermal stratification $S > 0$, as obtained by Shapiro and Fedorovich (2004). In Figure 8.2, an increase in Gr and Gc produces an increase in the velocity due to the enhanced mass and thermal buoyancy effects. Figure 8.3, shows the velocity of the fluid at different time t. The fluid velocity increases with time until it reaches a steady condition.

Figure 8.5, illustrates how S and F influences the temperature profile. The temperature is observed to fall as both the mass and thermal stratification parameters are increased, with the thermal stratification parameter showing a bigger decrease

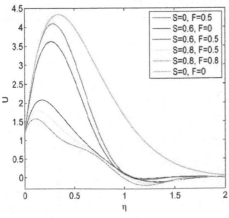

Figure 8.1 Impacts of S and F on velocity for $Gr = 5$, $Gc = 5$, $t = 1.9$, $a = 0.1$

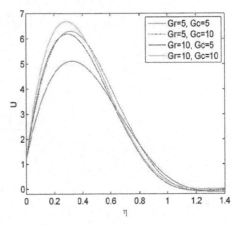

Figure 8.2 Impacts of Gr and Gc on velocity profile for $S = 0.2$, $F = 0.1$, $t = 1.9$, $a = 0.1$

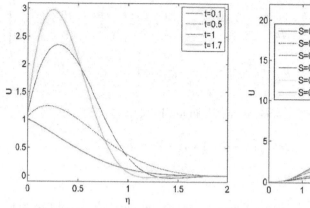

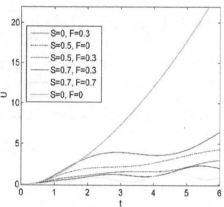

Figure 8.3 Velocity profile at different time for $Gr = 5$, $Gc = 5$, $S = 0.5$, $F = 0.3$, $a = 0.1$

Figure 8.4 Impacts of S and F on velocity against time for $Gr = 5$, $Gc = 5$, $a = 0.1$, $y = 1.8$

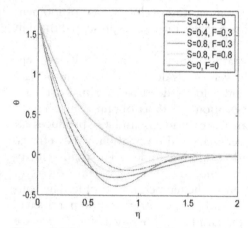

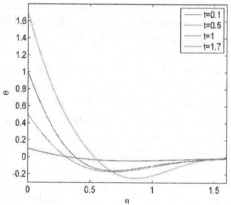

Figure 8.5 Impacts of S and F on temperature for $Gr = 5$, $Gc = 5$, $t = 1.9$, $a = 0.1$

Figure 8.6 Temperature profile at different time for $Gr = 5$, $Gc = 5$, $S = 0.5$, $F = 0.3$, $a = 0.1$

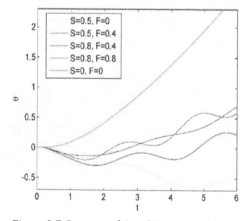

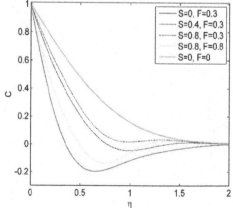

Figure 8.7 Impacts of S and F on temperature against time for $Gr = 5$, $Gc = 5$, $a = 0.1$, $y = 1.7$

Figure 8.8 Impacts of S and F on concentration for $Gr = 5$, $Gc = 5$, $t = 1.7$, $a = 0.1$

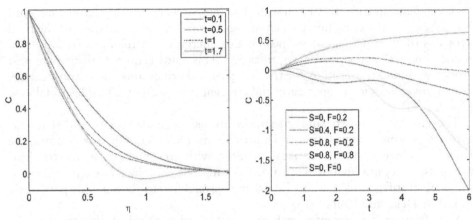

Figure 8.9 Concentration profile at different time for *Gr* = 5, *Gc* = 5, *S* = 0.5, *F* = 0.3, *a* = 0.1

Figure 8.10 Impacts of *S* and *F* on concentration against time for *Gr* = 5, *Gc* = 5, *a* = 0.1, *y* = 1.7

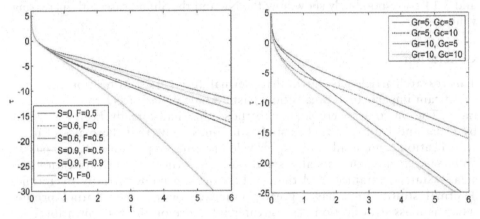

Figure 8.11 Impacts of *S* and *F* on skin friction for *Gr* = 5, *Gc* = 5, *a* = 0.1

Figure 8.12 Impacts of *Gr* and *Gc* on skin friction for *S* = 0.5, *F* = 0.3, *a* = 0.1

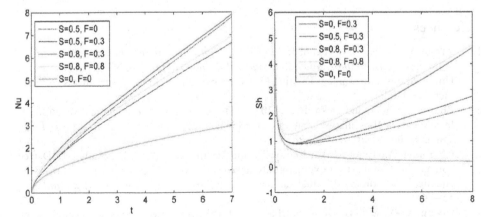

Figure 8.13 Impacts of *S* and *F* on Nu for *Gr* = 5, *Gc* = 5, *a* = 0.1

Figure 8.14 Impacts of S and F on Sh for *Gr* = 5, *Gc* = 5, *a* = 0.1

than the mass stratification parameter. The temperature increases with time until it reaches a steady condition as seen in Figure 8.6. The concentration of the fluid drops as the thermal and mass stratification increase, but the mass stratification reduces more than the thermal stratification. This study is more realistic than past works without stratification because it applies thermal and mass stratification, which lower velocity, temperature, and concentrations relative to the special case $(S = 0, F = 0)$.

The effect of the two kinds of stratifications on fluid velocity and temperature are presented against time in Figures 8.4 and 8.7. Without stratification, the velocity and temperature increase indefinitely with time, however when stratification exists, they gradually achieve a stable state. The combined impacts of the two kinds of stratifications result in a quicker approach to steady states, as previously found by Deka and Paul (2013).

Skin friction is represented in Figures 8.11 and 8.12 which demonstrate the impacts of S, F, Gr, Gc, respectively. Initially, skin friction reduces with time. It increases with S and F but declines with an increase in Gr, Gc. Figures 8.13 and 8.14 correspondingly show how the Nu and the Sh increase with increasing values of t.

Conclusion

This research article has shown that thermal and mass stratification can have significant impacts on unsteady flow past a vertical plate accelerates exponentially. The outcomes were similar to those that had already been published by Deka and Paul (2013). The study's findings show that thermal and mass stratification can considerably lower velocity and temperature as time passes and the system reaches a stable state. But, when there is no stratification, no steady state is attained, and these values decline monotonically over time. Thermal stratification has a greater influence on the temperature profile, whereas mass stratification has a greater influence on the concentration profile. The results of this research will be beneficial in the design of different systems with stratification.

References

1.   Abramowitz, M. and Stegun, I. A. (1972). *Handbook of Mathematical Functions With Formulas, Graphs, and Mathematical Tables* [Report]. UNT Digital Library; United States. Government Printing Office. https://digital.library.unt.edu/ark:/67531/metadc40302/.

2.   Bhattacharya, A. and Deka, R. K. (2011). Theoretical study of chemical reaction effects on vertical oscillating plate immersed in a stably stratified fluid. *Res. J. Appl. Sci. Engg. Technol.*, 3(09), 887–898.

3.   Chamkha, A., Raju, M. C., Reddy, T., and Varma, S. (2016). Unsteady MHD free convection flow past an exponentially accelerated vertical plate with mass transfer, chemical reaction and thermal radiation. *Dev. Res. Micro. Nano. Therm. Fluid Sci.*, 55–74.

4.   Cheng, C.-Y. (2008). Double-diffusive natural convection along a vertical wavy truncated cone in non-Newtonian fluid saturated porous media with thermal and

mass stratification. *Int. Comm. Heat Mass Trans.*, 35(8), 985–990. https://doi.org/10.1016/j.icheatmasstransfer.2008.04.007.

5.   Cheng, C.-Y. (2009). Combined heat and mass transfer in natural convection flow from a vertical wavy surface in a power-law fluid saturated porous medium with thermal and mass stratification. *Int. Comm. Heat Mass Trans.*, 36(4), 351–356. https://doi.org/10.1016/j.icheatmasstransfer.2009.01.003.

6.   Das, U. N., Deka, R., and Soundalgekar, V. M. (1994). Effects of mass transfer on flow past an impulsively started infinite vertical plate with constant heat flux and chemical reaction. *Forschung Im Ingenieurwesen*, 60(10), 284–287. https://doi.org/10.1007/BF02601318.

7.   Deka, R. K. and Paul, A. (2013). Convectively driven flow past an infinite moving vertical cylinder with thermal and mass stratification. *Pramana*, 81(4), 641–665. https://doi.org/10.1007/s12043-013-0604-6.

8.   Hetnarski, R. B. (1975). An algorithm for generating some inverse Laplace transforms of exponential form. *Zeitschrift Für Angewandte Mathematik Und Physik ZAMP*, 26(2), 249–253. https://doi.org/10.1007/BF01591514.

9.   Kalita, N., Deka, R. K., and Nath, R. S. (2023). Unsteady flow past an accelerated vertical plate with variable temperature in presence of thermal stratification and chemical reaction. *East Eur. J. Phy.*, 3, Article 3. https://doi.org/10.26565/2312-4334-2023-3-49.

10.   Magyari, E., Pop, I., and Keller, B. (2006). Unsteady free convection along an infinite vertical flat plate embedded in a stably stratified fluid-saturated porous medium. *Trans. Porous Media*, 62(2), 233–249. https://doi.org/10.1007/s11242-005-1292-6.

11.   Nath, R. S. and Deka, R. K. (2023). The effects of thermal stratification on flow past an infinite vertical plate in presence of chemical reaction. *East Eur. J. Phy.*, 3, Article 3. https://doi.org/10.26565/2312-4334-2023-3-19.

12.   Park, J. S. (2001). Transient buoyant flows of a stratified fluid in a vertical channel. *KSME Int. J.*, 15(5), 656–664. https://doi.org/10.1007/BF03184382.

13.   Park, J. S. and Hyun, J. M. (1998). Technical note transient behavior of vertical buoyancy layer in a stratified fluid. *Int. J. Heat Mass Trans.*, 41(24), 4393–4397. https://doi.org/10.1016/S0017-9310(98)00175-6.

14.   Paul, A. and Deka, R. K. (2017). Unsteady natural convection flow past an infinite cylinder with thermal and mass stratification. *Int. J. Engg. Math.*, 2017, e8410691. https://doi.org/10.1155/2017/8410691.

15.   Ramakrishna, S. B., N. L., R., and Kumar Thavada, S. (2022). Effects of chemical reaction, Soret and Lorentz force on Casson fluid flow past an exponentially accelerated vertical plate: A comprehensive analysis. *Heat Transfer*, 51(2), 2237–2257. https://doi.org/10.1002/htj.22398.

16.   Shapiro, A. and Fedorovich, E. (2004). Unsteady convectively driven flow along a vertical plate immersed in a stably stratified fluid. *J Fluid Mec.*, 498, 333–352. https://doi.org/10.1017/S0022112003006803.

17.   Vemula, R., Vijaya, S., and Varma, S. (2006). Radiation and mass transfer effects on MHD free convection flow past an exponentially accelerated vertical plate with variable temperature. *J Engg. Appl. Sci.*, 4.

9 Thermal and mass stratification effects on unsteady flow past an accelerated infinite vertical plate with exponential temperature variation and variable mass diffusion

Himangshu Kumar[a], Rudra Kanta Deka and Rupam Shankar Nath

Department of Mathematics, Gauhati University, India

Abstract

This study investigates how mass and thermal stratification affects the unstable flow past an accelerating infinite perpendicular plate with exponential temperature variation and variable mass diffusion. The method of the Laplace transformation is applied to solve the governing equations for unit Prandtl and Schmidt numbers. The Nusselt number, Sherwood number, and skin friction are derived and displayed in graphs. The outcomes of this study will provide a useful understanding of the interaction between thermal and mass stratification and their influence on unsteady flow characteristics. The results obtained can be useful in the design of heat exchangers and other engineering applications.

Keywords: Thermal stratification, mass stratification, unsteady flow, exponential temperature variation, accelerated plate

Introduction

The investigation of unsteady flow through an accelerating endless vertical plate has drawn the interest of investigators for decades because of its vast variety of practical applications. The influences of temperature and mass stratification on flow have been studied widely. Thermal stratification is the phenomenon of temperature variations between fluid layers and mass stratification is the phenomenon of variances in mass concentration that exist between layers of a fluid. These stratification effects can have a significant impact on the flow through an endless perpendicular plate. The current investigation, which is inspired by prior research, seeks to investigate the consequence of mass and thermal stratification, as well as exponential temperature variation and varying mass diffusion, on an unstable flow passing an infinitely fast perpendicular plate. These studies have shown that stratification may produce a wide range of flow characteristics and have a major impact on flow dynamics.

[a]himangshukumar307@gmail.com

Literature review

Many researchers have explored unstable flow across an exponentially accelerating infinite vertical plate theoretically and empirically (Rajesh, 2009; Kumar, 2011, 2012; Asogwa, 2012; Lakshmi). Rajput (2012) exploresd how radiation affected a suddenly launched perpendicular plate with variable mass and heat exchange rates for magnetohydrodynamic flow. Uwanta (2012) conducted research to evaluate the effects of mass and heat transmission with temperature fluctuations and exponential mass movement. The impact of chemical reactions and emission on a non-steady magnetohydrodynamic flow, whenever the temperature and mass transmission of a vertically accelerating porous plate vary, was examined. Babu (2017) and Sivakumar studied the influence of drag coefficient, the mass flow rate, and heat flow rate on parabolic flow through an endless perpendicular plate under the conditions of changing mass diffusion and thermal radiation.

Nomenclature

$\alpha \rightarrow$ Thermal diffusivity

$\beta \rightarrow$ Volumetric thermal expansion coefficient

$\beta^* \rightarrow$ Concentration-dependent volumetric expansion coefficient

$\gamma \rightarrow$ Thermal stratification parameter

$\nu \rightarrow$ Kinematic viscosity

$\tau \rightarrow$ Skin-friction without dimension

$\theta \rightarrow$ Temperature without a dimension

$\xi \rightarrow$ Mass stratification parameter

A, a, a' $\rightarrow$ Constant

$C \rightarrow$ Non-dimensional concentration

$C' \rightarrow$ Species fluid concentration

$C'_\infty \rightarrow$ Concentration of the fluid distant from the plate

$C'_w \rightarrow$ Plate's concentration

$D \rightarrow$ Coefficient of mass diffusion

$F \rightarrow$ Non-dimensional mass stratification parameter

$g \rightarrow$ Gravitational acceleration

$Gc \rightarrow$ Mass Grashof number

$Gr \rightarrow$ Thermal Grashof number

$Pr \rightarrow$ Prandtl number

$Sc \rightarrow$ Schmidt number

$S \rightarrow$ Non-dimensional thermal stratification parameter

$t \rightarrow$ Time with no dimensions

$T' \rightarrow$ Level of heat in the fluid

$t' \rightarrow$ Time

$T'_\infty \rightarrow$ Temperature of the fluid distant from the plate

$T'_w \rightarrow$ Plate's temperature

$U \rightarrow$ Velocity without dimensions

$u' \rightarrow$ The fluid's velocity along the x'-axis

$u_0 \rightarrow$ Acceleration of the plate

y → Dimensionless coordinate that is perpendicular to the plate
y' → Co-ordinate that is perpendicular to the plate

Mathematical analysis

An unstable flow across an accelerating upright plate with exponential tempera-ture change and fluctuating mass diffusion is described in Cartesian co-ordinates (x', y'). In this case, the endless plate is at right angles to the y' axis, and the x' axis is aligned to it. In the beginning, both the fluid and the surface of the plate are at the same initial temperature, T'$_\infty$, and concentration, C'$_\infty$, everywhere. At the instant t', the plate initiated acceleration within its plane, exhibiting a veloc-ity denoted as $u' = u_o t'$. The plate concentration increased straight with time t', while the temperature level close to the plate reached a value $T'_\infty + (T'_W - T'_\infty)e^{a't'}$. Hence, the unsteady flow can be described by the following equations using the standard Boussinesq approximation:

$$\frac{\partial u'}{\partial t'} = g\beta(T' - T'_\infty) + g\beta^*(C' - C'_\infty) + v\frac{\partial^2 u'}{\partial y'^2} \tag{1}$$

$$\frac{\partial T'}{\partial t'} = \alpha\frac{\partial^2 T'}{\partial y'^2} - \gamma u' \tag{2}$$

$$\frac{\partial c'}{\partial t'} = D\frac{\partial^2 c'}{\partial y'^2} - \xi u' \tag{3}$$

Boundary and starting conditions are as follows:

$$u' = 0 ; \qquad T' = T'_\infty ; \qquad C' = C'_\infty; \qquad \forall\, y', t' \leq 0$$

$$u' = u_0 t'; \quad T' = T'_\infty + (T'_W - T'_\infty)e^{a't'}; \quad C' = C'_\infty + (C'_W - C'_\infty)At'; \quad at, y' = 0, t' > 0 \tag{4}$$

$$u' = 0 ; \qquad T' \rightarrow T'_\infty ; \qquad C' \rightarrow C'_\infty ; \qquad at, y' \rightarrow \infty, t' > 0$$

Introducing the quantities without dimensions:

$$U = \frac{u'}{(u_0 v)^{\frac{1}{3}}} ; \quad t = \frac{t'u_0^{\frac{2}{3}}}{v^{\frac{1}{3}}} ; \quad y = \frac{y'u_0^{\frac{1}{3}}}{v^{\frac{2}{3}}} ; \quad \theta = \frac{T' - T'_\infty}{T'_W - T'_\infty} ; \quad C = \frac{C' - C'_\infty}{C'_W - C'_\infty} ; \quad Gr = \frac{g\beta(T'_W - T'_\infty)}{u_0}; \quad Pr = \frac{v}{\alpha} ;$$

$$Gr = \frac{g\beta(T'_W - T'_\infty)}{u_0} ; \quad Pr = \frac{v}{\alpha} ; \quad Sc = \frac{v}{D} ; \quad Gc = \frac{g\beta^*(C'_W - C'_\infty)}{u_0} ; \quad a = \frac{v^{\frac{1}{3}}a'}{u_0^{\frac{2}{3}}} ; \quad S = \frac{\gamma v^{\frac{2}{3}}}{u_0^{\frac{1}{3}}(T'_W - T'_\infty)} ;$$

$$F = \frac{\xi v^{\frac{2}{3}}}{u_0^{\frac{1}{3}}(C'_W - C'_\infty)}$$

where $A = \left(\frac{u_0^2}{v}\right)^{\frac{1}{3}}$ is a constant.

   Equations (1–3) reduces to,

$$\frac{\partial U}{\partial t} = \frac{\partial^2 U}{\partial y^2} + Gr.\theta + Gc.C \tag{5}$$

$$\frac{\partial \theta}{\partial t} = \frac{1}{Pr}\frac{\partial^2 \theta}{\partial y^2} - S.U \tag{6}$$

$$\frac{\partial C}{\partial t} = \frac{1}{Sc}\frac{\partial^2 C}{\partial y^2} - F.U \tag{7}$$

Boundary and initial condition in the absence of dimensions are:

$$U = 0; \quad \theta = 0; \quad C = 0; \qquad \forall\, y, t \leq 0$$

$$U = t; \quad \theta = e^{at}; \quad C = t; \quad \text{at}, y = 0, t > 0 \tag{8}$$

$$U = 0; \quad \theta \to 0; \quad C \to 0; \quad \text{as}, \ y \to \infty, t > 0$$

Method of solution

Using (8) and the Laplace transformation technique, we obtained the equations representing the temperature distribution, concentration profile, and velocity profile by solving the equations (5–7) for Pr = Sc = 1 and the expressions are as follows:

$$U = \tfrac{1}{2}\{K_1(iQ) + K_1(-iQ)\} + \tfrac{Gr}{2iQ}\{K_2(a,-iQ) - K_2(a,iQ)\} + \tfrac{Gc}{2iQ}\{K_1(-iQ) - K_1(iQ)\} \tag{9}$$

$$\theta = \tfrac{F.Gc}{Q^2}\left[\tfrac{e^{at}}{2}\left\{e^{-y\sqrt{a}}erfc\left(\tfrac{y}{2\sqrt{t}}-\sqrt{at}\right) + e^{y\sqrt{a}}erfc\left(\tfrac{y}{2\sqrt{t}}+\sqrt{at}\right)\right\}\right] - \tfrac{S.Gc}{Q^2}\left\{\left(t+\tfrac{y^2}{2}\right)erfc\left(\tfrac{y}{2\sqrt{t}}\right) - y\sqrt{\tfrac{t}{\pi}}e^{-\tfrac{y^2}{4t}}\right\} +$$
$$\tfrac{S}{2iQ}\{k_1(iQ) - K_1(-iQ)\} + \tfrac{S.Gr}{2Q^2}\{K_2(a,-iQ) + K_2(a,iQ)\} + \tfrac{S.Gc}{2Q^2}\{K_1(-iQ) + K_1(iQ)\} \tag{10}$$

$$C = \tfrac{S.Gr}{Q^2}\left\{\left(t+\tfrac{y^2}{2}\right)erfc\left(\tfrac{y}{2\sqrt{t}}\right) - y\sqrt{\tfrac{t}{\pi}}e^{-\tfrac{y^2}{4t}}\right\} - \tfrac{F.Gr}{Q^2}\left[\tfrac{e^{at}}{2}\left\{e^{-y\sqrt{a}}erfc\left(\tfrac{y}{2\sqrt{t}}-\sqrt{at}\right) + e^{y\sqrt{a}}erfc\left(\tfrac{y}{2\sqrt{t}}+\sqrt{at}\right)\right\}\right] +$$
$$\tfrac{F}{2iQ}\{k_1(iQ) - K_1(-iQ)\} + \tfrac{F.Gr}{2Q^2}\{K_2(a,-iQ) + K_2(a,iQ)\} + \tfrac{F.Gc}{2Q^2}\{K_1(-iQ) + K_1(iQ)\} \tag{11}$$

where, $Q = \sqrt{S.Gr + F.Gc}$
and K_i's are inverse Laplace transforms given by

$$K_1(iq) = L^{-1}\left\{\frac{e^{-y\sqrt{s+iq}}}{s^2}\right\}$$

$$K_2(b, iq) = L^{-1}\left\{\frac{e^{-y\sqrt{s+iq}}}{s-b}\right\}$$

Nusselt number, skin friction, and Sherwood number

Non-dimensional determinations of the plate's Nusselt number (related to heat transfer), Sherwood number (related to mass transfer), and skin friction (relative to momentum transfer) is as follows:

$$Nu = -\frac{d\theta}{dy}\bigg|_{y=0} \quad ; \quad Sh = -\frac{dC}{dy}\bigg|_{y=0} \quad ; \quad \tau = -\frac{dU}{dy}\bigg|_{y=0}$$

So, using the expression in equations (9), (10) and (11) we get,

$$Nu = \frac{F.Gc}{Q^2}\left\{\sqrt{a}e^{at}erf(\sqrt{at}) + \frac{1}{\sqrt{\pi t}}\right\} - \frac{S.Gc}{Q^2}\left(2\sqrt{\frac{t}{\pi}}\right) + \frac{S}{Q}\left[\left\{t\sqrt{\frac{Q}{2}}(z_1 + z_2) - \frac{\sin(Qt)}{\sqrt{\pi t}}t - \frac{(z_1 - z_2)}{4\sqrt{Q}}\right\} + \frac{Gr}{Q}\left\{e^{at}(z_3 z_5 - z_4 z_6) + \frac{\cos(Qt)}{\sqrt{\pi t}}\right\} + \frac{Gc}{Q}\left\{t\sqrt{\frac{Q}{2}}(z_1 - z_2) + \frac{\cos(Qt)}{\sqrt{\pi t}}t + \frac{(z_1 + z_2)}{2\sqrt{2Q}}\right\}\right] \tag{12}$$

$$Sh = \frac{S.Gr}{Q^2}\left(2\sqrt{\frac{t}{\pi}}\right) - \frac{F.Gr}{Q^2}\left\{\sqrt{a}e^{at}erf(\sqrt{at}) + \frac{1}{\sqrt{\pi t}}\right\} + \frac{F}{Q}\left[\left\{t\sqrt{\frac{Q}{2}}(z_1 + z_2) - \frac{\sin(Qt)}{\sqrt{\pi t}}t - \frac{(z_1 - z_2)}{4\sqrt{Q}}\right\} + \frac{Gr}{Q}\left\{e^{at}(z_3 z_5 - z_4 z_6) + \frac{\cos(Qt)}{\sqrt{\pi t}}\right\} + \frac{Gc}{Q}\left\{t\sqrt{\frac{Q}{2}}(z_1 - z_2) + \frac{\cos(Qt)}{\sqrt{\pi t}}t + \frac{(z_1 + z_2)}{2\sqrt{2Q}}\right\}\right] \tag{13}$$

$$\tau = t\sqrt{\frac{Q}{2}}(z_1 - z_2) + t\frac{\cos(Qt)}{\sqrt{\pi t}} + \frac{(z_1 + z_2)}{2\sqrt{2Q}} + \frac{Gr}{Q}\left\{\frac{\sin(Qt)}{\sqrt{\pi t}} - e^{at}(z_3 z_6 + z_4 z_5)\right\} + \frac{Gc}{Q}\left[t\frac{\sin(Qt)}{\sqrt{\pi t}} + \frac{(z_1 - z_2)}{2\sqrt{2Q}} - t\sqrt{\frac{Q}{2}}(z_1 + z_2)\right] \tag{14}$$

where,

$$erf(\sqrt{iQt}) = z_1 + iz_2$$

$$\sqrt{a + iQ} = z_3 + iz_4$$

$$erf(\sqrt{(a + iQ)t}) = z_5 + iz_6$$

Classical solution (S=0; F=0)
For the classical solution, we first put $\gamma = 0$ in equation (2) and $\xi = 0$ in equation (3). After that, they are non-dimensionalized by using the same set of dimensionless quantities. Thus the expressions of temperature, concentration, and velocity are attained as follows:

$$\theta^{cs} = \frac{e^{at}}{2}\left\{e^{-y\sqrt{a}}erfc\left(\frac{y}{2\sqrt{t}} - \sqrt{at}\right) + e^{y\sqrt{a}}erfc\left(\frac{y}{2\sqrt{t}} + \sqrt{at}\right)\right\} \tag{15}$$

$$C^{cs} = \left(t + \frac{y^2}{2}\right)erfc\left(\frac{y}{2\sqrt{t}}\right) - y\sqrt{\frac{t}{\pi}}e^{-\frac{y^2}{4t}} \tag{16}$$

and

$$U^{cs} = t\left\{\left(1 + \frac{y^2}{2t}\right)erfc\left(\frac{y}{2\sqrt{t}}\right) - \frac{y}{\sqrt{\pi t}}e^{-\frac{y^2}{4t}}\right\} + \frac{y.Gr}{2}\left[\frac{e^{at}}{2\sqrt{a}}\left\{e^{-y\sqrt{a}}erfc\left(\frac{y}{2\sqrt{t}} - \sqrt{at}\right) - e^{y\sqrt{a}}erfc\left(\frac{y}{2\sqrt{t}} + \sqrt{at}\right)\right\}\right] + \frac{y.Gc}{2}\cdot\frac{t^{\frac{3}{2}}}{3}\left[\frac{4}{\sqrt{\pi}}\left(1 + \frac{y^2}{4t}\right)e^{-\frac{y^2}{4t}} - \frac{y}{2\sqrt{t}}\left(6 + \frac{y^2}{t}\right)erfc\left(\frac{y}{2\sqrt{t}}\right)\right] \tag{17}$$

In the classical scenario, the rate of momentum transmission, transfer of heat, and mass exchange rate are derived and can be expressed as follows:

$$\tau^{cs} = -\frac{dU^{cs}}{dy}\bigg|_{y=0} = 2\sqrt{\frac{t}{\pi}} - \frac{Gr.e^{at}}{2\sqrt{a}}erf(\sqrt{at}) - \frac{2.Gc.t^{\frac{3}{2}}}{3\sqrt{\pi}}$$

$$Nu^{cs} = -\frac{d\theta^{cs}}{dy}\bigg|_{y=0} = e^{at}\sqrt{a}.erf(\sqrt{at}) + \frac{1}{\sqrt{\pi t}} \quad \text{and} \quad Sh^{cs} = -\frac{dC^{cs}}{dy}\bigg|_{y=0} = 2\sqrt{\frac{t}{\pi}}$$

Results and discussion

The outcomes derived from the numerical computations were displayed in diagrams to explain the physical significance of the situation. In the graphs, the relationships between the physical components Gr, Gc, S, F, and time t were shown, together with variations in concentration, temperature, velocity, rate of momentum transmission, transfer of heat, and mass exchange rate.

Figures 9.1–9.3 shows the velocity profile with and without stratification depending on the values of S, F, Gr, Gc, a, and t. The velocity of a stratified fluid is seen to be lower than the velocity of an unstratified fluid. The velocity decreases with the increment of mass and thermal stratification parameter (S and F). The drop in speed is because of the stacking effect of stratification of heat, which behaves like a frictional force. Increasing the thermal Grashof number (Gr) results in a reduction in velocity, whereas increasing the mass Grashof number (Gc) causes an increase in speed. The buoyancy force grows as the thermal Grashof number rises, and this flotational impulse operates against the flow to lower its velocity. It is obvious from Figure 9.3 that the classical velocity will continue to grow without bounds as time passes; but, when stratifications are present, the velocity will eventually attain a steady state. When more time passes, the combination of the two stratifications causes a more rapid convergence to the steady state.

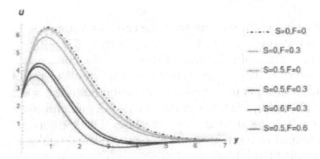

Figure 9.1 Influences of S and F on velocity for Gr = 3, Gc = 6, a = 0.1, t = 2.5

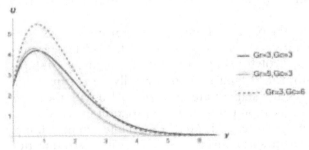

Figure 9.2 Influences of Gr and Gc on velocity for S = 0.3, F = 0.2, a = 0.1, t = 2.5

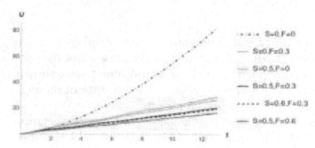

Figure 9.3 Influences of S and F on velocity against time for Gr=3, Gc=6, y=0.8, a=0.1

Figure 9.4 Influences of S and F on temperature for Gr = 3, Gc = 6, t = 2.5, a = 0.1

Figures 9.4–9.8 displays the temperature along with the concentration profile as a function of y for different quantities of S, F, Gr, Gc, a, and t. It is noticed from the diagrams that at the plate, both the temperature and the concentration are at their peaks, and later on they are getting closer and closer to zero. With the increment of Gr and Gc, temperature and concentration both decrease. Temperature goes down when S goes up, but it goes up when F goes up and concentration decreases as F increases but it increases as S increases. Furthermore, it has been found that the concentration and temperature of the stratified fluid are lower than those of the unstratified fluid. How the different stratifications affect the concentration and temperature of the fluid over time is seen in Figures 9.6 and 9.9, respectively. These two figures conclude that the classical concentration and temperature grow unrestrictedly across time, but with stratifications, they approach a stable state.

Figures 9.10–9.12 illustrates the pattern of momentum transfer rate, Nusselt number, and mass exchange rate over time for various values of temperature and mass stratification factors, involving the traditional case. Figure 9.10 shows that the skin friction for unstratified fluid is monotonically decreasing while for the stratified fluid, it arrives at a stable state over time. Also with the increment of mass and thermal stratification parameter, skin friction increases. Skin friction is created by the plate and the fluid interaction. Fluid temperature or density differs more from the plate surface as the thermal or mass stratification parameter increases. Due to the changing temperature or density of the fluid, the fluid and plate's surface interact more forcefully thereby increasing skin friction. Similarly from Figure 9.11, it is clear that the transmission of heat goes down for a short period and after that, it increases. Also with the increment in the thermal stratification parameter Nusselt number goes up but it declines as the mass stratification parameter increases. Similarly, from Figure 9.12 it is clear that the mass transfer rate goes up with time. It likewise goes up when the mass stratification parameter goes up, but it goes down as the thermal stratification value goes up.

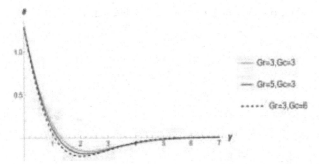

Figure 9.5 Influences of Gr and Gc on temperature for S = 0.3, F = 0.2, a = 0.1, t = 2.5

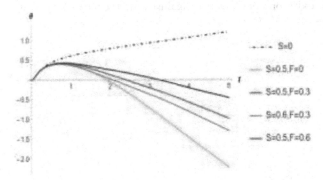

Figure 9.6 Influences of S and F on temperature against time for Gr = 3, Gc = 6, y = 0.8, a = 0.1

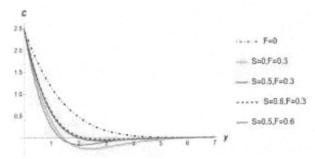

Figure 9.7 Influences of S and F on concentration for Gr = 3, Gc = 6, t = 2.5, a = 0.1

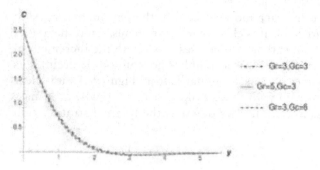

Figure 9.8 Influences of Gr and Gc on concentration for S = 0.3, F = 0.2, a = 0.1, t = 2.5

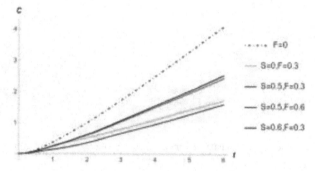

Figure 9.9 Influences of S and F on concentration against time for Gr = 3, Gc = 6, y = 0.8, a = 0.1

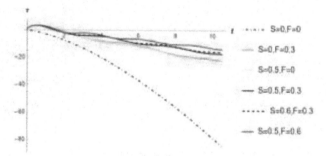

Figure 9.10 Influences of S and F on skin friction for Gr = 3, Gc = 6, a = 0.1

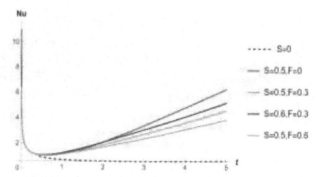

Figure 9.11 Influences of S and F on Nusselt number for Gr = 3, Gc = 6, a = 0.1

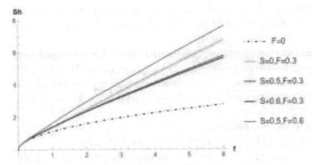

Figure 9.12 Influences of S and F on Sherwood number for Gr = 3, Gc = 6, a = 0.1

Conclusion

Based on the results derived from the preceding discussion, the following are the conclusions of this study:

(i) Velocity falls off as S, F, and Gr increase, but increases as Gc increases. The fluid that has undergone thermal and mass stratification has a lower velocity than the unstratified fluid.

(ii) When S, Gr, and Gc are increased, the temperature drops. However, when F rises, so does the temperature. Compared to non-stratified fluid, stratified fluid has a lower temperature. Temperature is highest close to the plate.

(iii) Increases in F, Gr, and Gc result in lower concentrations, while increases in S result in higher concentrations. Fluid with a stratification has a lower concentration than one without a stratification. Concentration is maximum near the plate.

(iv) As time passes, it is shown that skin friction for stratified fluid approaches a fixed value, whereas, in the classical situation, these values drop monotonically. It also increases as S and F increase.

(v) Nusselt number initially falls for a short period but before it increases. It also goes up as S goes up, but it goes down as F goes up.

(vi) The value of Sherwood's number grows as time passes. Also, with the increment of F Sherwood number increases but it decreases as S increases.

References

1. Abramowitz, M. S. (1988). Handbook of mathematical functions with formulas, graphs, and mathematical tables. American Association of Physics Teachers. 55, 1–1045.

2. Asogwa, K. U. (2012). Flow past an exponentially accelerated infinite vertical plate and temperature with variable mass diffusion. *Int. J. Comp. Appl.*, 45(2), 1–7.

3. Babu, V. S. (2017). Radiation and chemical reaction effects on unsteady MHD free convection flow past a linearly accelerated vertical porous plate with variable temperature and mass diffusion. *Global J. Pure Appl. Math.*, 13(9), 5341–5358.

4. Balla, C. S. (2015). Radiation effects on unsteady MHD convective heat and mass transfer past a vertical plate with chemical reaction and viscous dissipation. *Alexandria Engg. J.*, 54(3), 661–671.

5.   Bhattacharya, A. (2011). Theoretical study of chemical reaction effects on vertical oscillating plate immersed in a stably stratified fluid. *Res. J. Appl. Sci. Engg. Technol.*, 3(9), 887–898.

6.   Das, S. J. (2017). Natural convection near a moving vertical plate with ramped heat and mass fluxes in the presence of thermal radiation. *Defect Diff. For.*, 377, 211–232.

7.   Das, U. R. (1996). Mass transfer effects on flow past an impulsively started infinite vertical plate with constant mass flux—an exact solution. *Heat Mass Trans.*, 31(3), 163–167.

8.   Deka, R. K. (2009). Unsteady natural convection flows past an accelerated vertical plate in a thermally stratified fluid. *Theor. Appl. Mech.*, 36(4), 261–274.

9.   Deka, R. K. (2012). Mhd free convection flows past an impulsively started infinite vertical plate with thermal stratification and radiation. *ASME Int. Mech. Engg. Cong. Exp.*, 45233, 2665–2670.

10.   Deka, R. K. (2013). Convectively driven flow past an infinite moving vertical cylinder with thermal and mass stratification. *Pramana*, 81, 641–665.

11.   Ghosh, S. D. (2014). Transient mhd free convective flow of an optically thick gray gas past a moving vertical plate in the presence of thermal radiation and mass diffusion. *J. Appl. Fluid Mech.*, 8(1), 65–73.

12.   Hussanan, A. I. (2014). Unsteady boundary layer MHD free convection flow in a porous medium with constant mass diffusion and Newtonian heating. *Eur. Phy. J. Plus*, 129, 1–16.

13.   Iranian, D. L. (2015). Unsteady mhd natural convective flow over a vertical plate in thermally stratified media with variable viscosity and thermal conductivity. *Int. J. Comp. Appl.*, 121(3), 18–24.

14.   Kumar, A. V. (2011). Thermal diffusion and radiation effects on unsteady MHD flow past an impulsively started exponentially accelerated vertical plate with variable temperature and variable mass diffusion. *Int. J. Appl. Math. Anal. Appl.*, 6, 191–214.

15.   Kumar, A. V. (2012). Chemical reaction and radiation effects on MHD free convective flow past an exponentially accelerated vertical plate with variable temperature and variable mass diffusion. *Ann. Fac. Engg. Hunedoara*, 10(2), 195–202.

16.   Lakshmi, C. S. (n.d.). First-order chemical reaction effects on an exponentially accelerated isothermal vertical plate with variable mass diffusion in the presence of radiation. 8, 700–705.

17.   Manivannan, K. M. (2009). Radiation and chemical reaction effects on the isothermal vertical oscillating plate with variable mass diffusion. *Therm. Sci.*, 13(2), 155–162.

18.   Muthucumaraswamy, R. (2006). MHD and radiation effects on moving isothermal vertical plate with variable mass diffusion. *Theor. Appl. Mech.*, 33(1), 17–29.

19.   Muthucumaraswamy, R. (2013). Hydromagnetic flow past an exponentially accelerated isothermal vertical plate with uniform mass diffusion in the presence of chemical reaction of the first order. *Int. J. Appl. Mech. Engg.*, 18(1), 259–267.

20.   Paul, A. (2017). Unsteady natural convection flows past an infinite cylinder with thermal and mass stratification. *Int. J. Eng. Math.*, 8410691.

21.   Rajesh, V. (2009). Radiation and mass transfer effects on MHD free convection flow past an exponentially accelerated vertical plate with variable temperature. *ARPN J. Engg. Appl. Sci.*, 4(6), 20–26.

22.   Rajput, U. (2012). Radiation effects on mhd flow past an impulsively started vertical plate with variable heat and mass transfer. *Int. J. Appl. Math. Mech.*, 8(1), 66–85.

23.   Rath, C. (2023). The transient natural convective flow of a radiative viscous incompressible fluid past an exponentially accelerated porous plate with chemical reaction species. *Heat Trans.* 52(1), 467–494.

24. Richard, B. and Hetnarski, D. O. (1975). An algorithm for generating some inverse Laplace transforms of exponential form. *J. Appl. Math. Phy. (ZAMP)*, 26, 249–253.
25. Shah, N. A. (2019). Effects of double stratification and heat flux damping on convective flows over a vertical cylinder. *Chinese J. Phy.*, 60, 290–306.
26. Sivakumar, P. (n.d.). Investigation of skin friction and rate of heat and mass transfer effects on parabolic flow past an infinite vertical plate in the presence of thermal radiation with variable mass diffusion. XIII, 375–387.
27. Uwanta, I. (2012). Heat and mass transfer with variable temperature and exponential mass diffusion. *Int. J. Comput. Eng. Res.*, 2, 1487–1494.
28. Ziaei-Rad, M. S. (2016). Simulation and prediction of MHD dissipative nanofluid flow on a permeable stretching surface using artificial neural network. *Appl. Therm. Engg.*, 99, 373–382.

10 Hydro magnetic squeeze film between conducting transversely rough curved circular plates

Ms. Pragna A. Vadher[1,a], Gunamani B. Deheri[2], Akhil S. Mittal[3], Mahaveer P. Shekhawat[4] and Rakesh M. Patel[4]

[1]Principal, Government Science College, Idar – 383430, Gujarat, India

[2]Retired Professor of Mathematics, Department of Mathematics, Sardar Patel University, Vallabh Vidyanagar – 388120, Gujarat, India

[3]Department of Mathematics, Government Science College, Santrampur – 389260, Gujarat, India

[4]Department of Mathematics, Gujarat Arts and Science College, Ahmedabad – 380006, Gujarat, India

Abstract

The squeeze film among electrically conducting rough curved surfaces in the presence of a transverse hydro magnetic field is investigated. Bearing surfaces have rough transverse surfaces. Non-zero mean, variance, and skewness are the foundations for roughness. The Reynolds' expression is stochastically averaged with regards to the roughness parameters. The pressure distribution is calculated using appropriate feasible boundary conditions. Further, the expression of pressure is used to determine the load carrying capacity. Results are computed with the help of C programming and presented in tabular form. From these tables it is observed that roughness affects adversely. However, this adverse effect can be compensated to some extent by choosing the suitable combinations of conductivity, shape parameter and hydro magnetization. In addition, the positive effect of negative variance and negative skewness play a crucial role for the better performance of the bearing system. Moreover, it has been found that a bearing with a magnetic field may maintain a load even in the absence of a flow.

Keywords: Hydro magnetic lubrication, Reynolds' equation, transverse roughness, load profile

Introduction

It is common knowledge that after some use and wear, the bearing surfaces become rougher. Many writers have looked at the effects of surface irregularity, including Davies (1963), Burton [1963], Michell (1950), Tonder (1967, 1977), and others. C&T (1969.a, 1969.b, 1970), B&G (1973), T&S (1967), and Mishra et al. (2018) are other examples. In their extensive general study for the irregularity of both transverse and longitudinal surfaces, C&T (1969 a, b, 1970) made a number of conclusions. In various studies (Ting, 1975; Prakash and Tiwari,

[a]pragnavadher@rediffmail.com, [b]gm.deheri@rediffmail.com, [c]akhilsmittal@gmail.com, [d]mahaveer.maths@gmail.com, [e]rmpatel2711@gmail.com

1982; Prajapati, 1991, 1992; Guha, 1993; Gupta and Deheri, 1996; Andharia, et al., 1997, 1999), the analysis of the impact of surface irregularity was based on the C&T approach. In their studies of the performance of magnetic fluid-based squeeze films between annular and circular plates. Patel and Deheri (2003, 2004) also looked at the impact of surface roughness on bearing system performance.

While an increase in plate conductivity improves performance for circular plates (Prajapati, 1995), here it is demonstrated that system performance is negatively impacted by the bearing surfaces' transverse roughness. Naturally, the situation can be somewhat recovered in the event of negatively skewed roughness, especially when the variance is negative. Investigating the hydromagnetic squeeze film that develops among conducting transversely rough curved circular plates that are situated along surfaces determined by exponential and hyperbolic functions is the main objective of this research. Here, it is found that the performance of the bearing system is significantly enhanced by the curvature parameters of both plates.

Analysis and mathematical computations

Geometry which is used in this analysis is as under (Figure 10.1):

The description of roughness is based on C&T (1969.a; 1969.b; 1970).

The modified Reynolds' (1958) equation is obtained by reducing the equation of continuity and is expressed as

$$\nabla^2 p = \frac{dh/dt}{\left[\dfrac{2h^3}{\mu M^3}\left\{\left(\tanh\dfrac{M}{2}-\dfrac{M}{2}\right)-\dfrac{\psi h^3}{\mu C^2}\right\}\right]\left[\dfrac{\varphi_0+\varphi_1+1}{\varphi_0+\varphi_1+\dfrac{\tanh(M/2)}{M/2}}\right]} \tag{1}$$

where $\psi = \dfrac{KH}{h^3}$.

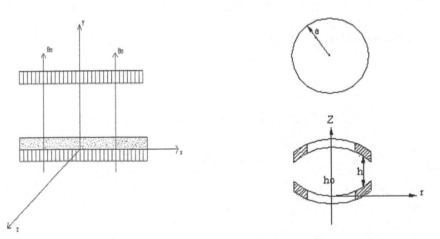

Figure 10.1 Hydro magnetic squeeze film bearing profile

According to the standard assumptions for hydro magnetic lubrication, the modified Reynolds' (1958) type equation for the lubricant film pressure is (Prajapati, 1995; Vadher et al., 2008).

$$\frac{1}{r}\frac{\partial}{\partial r}\left(r\frac{\partial p}{\partial r}\right) = \frac{\mu\dot{h}}{g(h)AB} \qquad (2)$$

where

$$g[h] = h^3 + 3h\sigma^2 + 3h^2\alpha + 3\alpha^2 h + 3\sigma^2\alpha + \alpha^3 + \varepsilon,$$

$$\bar{h} = \frac{h(r)}{h_0} = e\text{-}B^*R^2\text{-sec}(\text{-}C^*R^2)+1 \;,$$

$$A = \left[\frac{\psi}{c^2} - \frac{2}{M^3}\{\tanh(M/2)-(M/2)\}\right], \quad B = \left[\frac{\phi_0+\phi_1+1}{\phi_0+\phi_1+\dfrac{\tanh(M/2)}{(M/2)}}\right]$$

and

$$M = B_0 h\left(\frac{s}{\mu}\right)^{1/2}$$

Use of B.C.

$$p(a) = 0;$$

$$\frac{\partial p}{\partial r} = 0 \text{ at } r = 0 \qquad (3)$$

using non-dimensional terminology

$$B^* = Ba^{2,} \qquad\qquad C^* = Ca^2, \qquad\qquad R = \frac{r}{a},$$

$$\sigma^* = \frac{\sigma}{h_0}, \qquad\qquad \alpha^* = \frac{\alpha}{h_0}, \qquad\qquad \varepsilon^* = \frac{\varepsilon}{h_0^3}$$

$$g(\bar{h}) = \bar{h}^3 + \varepsilon^* + 3\sigma^{*2}\bar{h} + 3\bar{h}^2\alpha^* + 3\alpha^{*2}\bar{h} + 3\alpha^*\sigma^{*2} + \alpha^{*3},$$

$$\bar{h} = \frac{h(r)}{h_0} = e\text{-}B^*R^2\text{-sec}(\text{-}C^*R^2)+1$$

distributes the pressure in dimensionless form as

$$P = \frac{-ph^3}{\mu \dot{h} \pi a^2}$$

(4)

$$P = -\frac{1}{\pi AB} \int_R^1 \frac{R \frac{g(h)}{\int_R^1 \frac{R}{R} dR}}{R} dR$$

Then the L.C.C.given by

$$w = 2 * \pi \int_0^a p(r) \cdot r * dr$$

(5)

is obtained in non-dimensional form as

$$W = -\frac{wh^3}{\mu \dot{h} \pi^2 a^4}$$

(6)

$$W = -\frac{1}{\pi AB} \int_0^1 RP dR$$

Last but not least, the time it takes for the plate to move from the film thickness $h = h0$ at $t = t0$ to the film thickness $h = h1$ is determined as the non-dimensional squeezing time T from equation (4). $h = h_1$ at $t = t_1$.

$$\Delta T = \int_0^{t_1/t_0} \frac{Wh_0^2}{\mu \pi^2 a^4} dt$$

which means

$$\Delta T = \frac{1}{8\pi} I$$

(7)

where I is given by

$$I = -h_0^2 \int_1^{h_1/h_0} \frac{dh}{\left[\frac{2*A}{M^3} \left(\tanh \frac{M}{2} - \frac{M}{2} \right) - \frac{KH}{c^2} \right]} \cdot \left[\frac{1}{\phi_1 + \phi_0 + \frac{1 + \phi_0 + \phi_1}{\tanh(M/2)}} \right]$$

Values

M	4	56	8	10	12
$\Phi_0 + \Phi_1$	0	1	2	3	4
ψ	0	0.01	0.05	0.10	0.15
$\sigma*$	0	0.05	0.10	0.15	0.20
$\alpha*$	-0.1	-0.05	0	0.05	0.10
$\varepsilon*$	-0.1	-0.05	0	0.05	0.10
B*	-0.2	-0.1	0	0.1	0.2
C*	-0.2	-0.1	0	0.1	0.2

Results and discussion

The fluctuation in load with regard to the M is shown in Tables 10.1–10.7. For various values of $\phi_0 + \phi_1$, ψ, $\sigma*$, $\alpha*$, $\varepsilon*$, B* and C*, respectively.

These Tables show that the load increases significantly when viewed closely. The distribution of load w.r.t. $\phi_1 + \phi_0$ for different values of different parameters presented in Tables 10.8–10.13. According to these Tables, the load carrying capacity greatly rises in comparison to $\phi_0 + \phi_1$.

Table 10.1 Load of M and $\phi_0 + \phi_1$.

$\psi = 0.01$, $\sigma* = 0.05$, $\alpha* = -0.05$, $\varepsilon* = -0.05$, $\beta* = 0.1$,				C*= 0.1
M→				
$\Phi_0 + \Phi_1\downarrow$ 0.343100	0.395200	0.446700	0.491500	0.527000
0.527500	0.793200	1.117500	1.474600	1.844400
0.588900	0.925900	1.341000	1.802300	2.283500
0.619700	0.992300	1.452800	1.966200	2.503100
0.638100	1.032100	1.519900	2.064500	2.634900

Table 10.2 Load of M and ψ.

$\Phi_0 + \Phi_1 = 2$, $\sigma* = 0.05$, $\alpha* = -0.05$, $\varepsilon* = -0.05$, $\beta* = 0.1$,				C*= 0.1
M→				
$\psi\downarrow$ 0.624200	1.022600	1.562800	2.238900	3.048300
0.588900	0.925900	1.341000	1.802300	2.283500
0.480400	0.671900	0.855500	1.012500	1.139800
0.390500	0.500300	0.589000	0.654200	0.700900
0.328900	0.398500	0.449100	0.483200	0.506100

Table 10.3 Load of M and σ*.

$\Phi_0 + \Phi_1 = 2$, $\psi = 0.01$, $\alpha* = -0.05$, $\varepsilon* = -0.05$, $\beta* = 0.1$,				X*= 0.1
M→				
σ* ↓				
0.594100	0.934100	1.352900	1.818200	2.303700
0.588900	0.925900	1.341000	1.802300	2.283500
0.573900	0.902300	1.306700	1.756200	2.225100
0.550400	0.865400	1.253300	1.684400	2.134200
0.520600	0.818500	1.185500	1.593300	2.018700

Table 10.4 Load of M and α*.

$\Phi_0 + \Phi_1 = 2$, $\psi = 0.01$, $\sigma* = 0.05$, $\varepsilon* = -0.05$, $\beta* = 0.1$,				*C*= 0.1*
M→				
α* = ↓				
0.699500	1.099800	1.592800	2.140700	2.712300
0.588900	0.925900	1.341000	1.802300	2.283500
0.501000	0.787700	1.140700	1.533100	1.942500
0.430000	0.676100	0.979200	1.316000	1.667400
0.372100	0.585000	0.847200	1.138600	1.442700

Table 10.5 Load of M and ε*.

$\Phi_0 + \Phi_1 = 2$, $\psi = 0.01$, $\sigma* = 0.05$, $\alpha* = -0.05$, $\beta* = 0.1$,				*C*= 0.1*
M→				
ε* ↓				
0.627500	0.986500	1.428700	1.920200	2.432900
0.588900	0.925900	1.341000	1.802300	2.283500
0.554900	0.872400	1.263500	1.698100	2.151500
0.524500	0.824700	1.194400	1.605200	2.033800
0.497400	0.781900	1.132500	1.522000	1.928400

Table 10.6 Load of M and B*.

$\Phi_0 + \Phi_1 = 2$, $\psi = 0.01$, $\sigma* = 0.05$, $\alpha* = -0.05$, $\varepsilon* = -0.05$,				*C*= 0.1*
M→				
β* ↓				
0.383300	0.602700	0.872800	1.173100	1.486300
0.416900	0.655400	0.949200	1.275700	1.616300
0.485400	0.763100	1.105200	1.485400	1.882000
0.588900	0.925900	1.341000	1.802300	2.283500
0.727500	1.143800	1.656500	2.226400	2.820800

Table 10.7 Load of M and C*.

$\Phi_0+\Phi_1 = 2$, $\psi= 0.01$, $\sigma* = 0.05$, $\alpha* = -0.05$, $\varepsilon* = -0.05$, $\beta*=0.1$				
M→				
C* ↓ 1.198000	1.883600	2.727900	3.666300	4.645200
0.940400	1.478400	2.141200	2.877700	3.646100
0.737300	1.159200	1.678900	2.256400	2.858900
0.588900	0.925900	1.341000	1.802300	2.283500
0.495200	0.778600	1.127600	1.515500	1.920100

Table 10.8 Load of $\phi_0 + \phi_1$ and $\psi*$.

M = 8, $\sigma* = 0.05$, $\alpha* = -0.05$, $\varepsilon* = -0.05$, $\beta*=0.1$				C*=0.1
$\Phi_0+\Phi_1$→				
ψ ↓ 0.520600	1.302200	1.562800	1.693000	1.771200
0.446700	1.117500	1.341000	1.452800	1.519900
0.285000	0.712900	0.855500	0.926800	0.969600
0.196200	0.490800	0.589000	0.638100	0.667500
0.149600	0.374200	0.449100	0.486500	0.508900

Table 10.9 Load of $\phi_0 + \phi_1$ and $\sigma*$.

M = 8, $\psi= 0.01$, $\alpha* = -0.05$, $\varepsilon* = -0.05$, $\beta*=0.1$				C*=0.1
$\Phi_0+\Phi_1$→				
$\sigma*$ ↓ 0.450700	1.127300	1.352900	1.465600	1.533300
0.446700	1.117500	1.341000	1.452800	1.519900
0.435300	1.088900	1.306700	1.415700	1.481000
0.417500	1.044400	1.253300	1.357800	1.420500
0.394900	0.987800	1.185500	1.284300	1.343600

Table 10.10 Load of $\phi_0 + \phi_1$ and $\alpha*$.

M = 8, $\psi= 0.01$, $\sigma* = 0.05$, $\varepsilon* = -0.05$, $\beta*=0.1$				C*=0.1
$\Phi0+\Phi1$→				
$\alpha*$ ↓ 0.530600	1.327300	1.592800	1.725600	1.805200
0.446700	1.117500	1.341000	1.452800	1.519900
0.380000	0.950600	1.140700	1.235800	1.292900
0.326200	0.815900	0.979200	1.060800	1.109800
0.282200	0.706000	0.847200	0.917800	0.960200

Table 10.11 Load of $\phi_0 + \phi_1$ and $\varepsilon*$.

M = 8, ψ= 0.01, $\sigma*$ = 0.05, $\alpha*$ = −0.05, $\beta*$=0.1				$C*=0.1$
$\Phi 0+\Phi 1 \rightarrow$				
$\varepsilon*\downarrow$ 0.476000	1.190500	1.428700	1.547800	1.619300
0.446700	1.117500	1.341000	1.452800	1.519900
0.420900	1.052800	1.263500	1.368800	1.432000
0.397900	0.995300	1.194400	1.293900	1.353700
0.377300	0.943700	1.132500	1.226900	1.283500

Table 10.12 Load of $\phi_0 + \phi_1$ and $B*$.

M = 8, ψ = 0.01, $\sigma*$ = 0.05, $\alpha*$ = −0.05, $\varepsilon*$=−0.05				$C*=0.1$
$\Phi 0+\Phi 1 \rightarrow$				
$\beta*\downarrow$ 0.290800	0.727300	0.872800	0.945600	0.989300
0.316200	0.790900	0.949200	1.028300	1.075800
0.368200	0.921000	1.105200	1.197400	1.252600
0.446700	1.117500	1.341000	1.452800	1.519900
0.551900	1.380400	1.656500	1.794600	1.877500

Table 10.13 Load of $\phi_0 + \phi_1$ and $C*$.

M = 8, ψ= 0.01, $\sigma*$ = 0.05, $\alpha*$ = −0.05, $\varepsilon*$=−0.05, $\beta*$=0.1				
$\Phi 0+\Phi 1 \rightarrow$				
$C*\downarrow$ 0.908800	2.273100	2.727900	2.955300	3.091800
0.713300	1.784200	2.141200	2.319700	2.426800
0.559300	1.399000	1.678900	1.818800	1.902800
0.446700	1.117500	1.341000	1.452800	1.519900
0.375600	0.939600	1.127600	1.221600	1.278000

The load profile with different values of the porosity parameter is shown in Tables 10.14–10.18 below.

Additionally, the combined impact of the curvature parameters is significant. In Tables 10.19–10.27, the L.B.C. for various combinations of roughness characteristics, including variance, skewness, and standard deviation is shown.

The influence of curvature parameters is shown in Table 10.28 for various parameters with fixed values, such as: M, ψ , $\phi_0 + \phi_1$, $\sigma*$, $\varepsilon*$, $\alpha*$.

The aforementioned factor alone shows that in the event of rough bearings, the machine's lifespan may be increased.

Table 10.14 Load of ψ and σ*.

M = 8, $\Phi_0+\Phi_1= 2$, $\alpha* = -0.05$, $\beta* = 0.1$, $\varepsilon*=-0.05$				C*=0.1
ψ→				
σ*↓ 1.576500	1.352900	0.863100	0.594200	0.453000
1.562800	1.341000	0.855500	0.589000	0.449100
1.522800	1.306700	0.833600	0.573900	0.437600
1.460500	1.253300	0.799600	0.550400	0.419700
1.381500	1.185500	0.756300	0.520600	0.397000

Table 10.15 Load of ψ and α*.

M = 8, $\Phi_0+\Phi_1= 2$, $\sigma* = 0.05$, $\beta* = 0.1$, $\varepsilon*=-0.05$				C*=0.1
ψ→				
α*↓ 1.856200	1.592800	1.016100	0.699500	0.533400
1.562800	1.341000	0.855500	0.589000	0.449100
1.329400	1.140700	0.727700	0.501000	0.382000
1.141100	0.979200	0.624700	0.430000	0.327900
0.987300	0.847200	0.540500	0.372100	0.283700

Table 10.16 Load of ψ and ε*.

M = 8, $\Phi_0+\Phi_1= 2$, $\sigma* = 0.05$, $\alpha* = -0.05$, $\beta*=0.1$				C*=0.1
ψ→				
ε*↓ 1.665000	1.428700	0.911500	0.627500	0.478400
1.562800	1.341000	0.855500	0.589000	0.449100
1.472400	1.263500	0.806000	0.554900	0.423100
1.391900	1.194400	0.762000	0.524600	0.400000
1.319700	1.132500	0.722500	0.497400	0.379200

Table 10.17 Load of ψ and B*.

M = 8, $\Phi_0+\Phi_1= 2$, $\sigma* = 0.05$, $\alpha* = -0.05$, $\varepsilon*=-0.05$				C*=0.1
ψ→				
β*↓ 1.017200	0.872800	0.556800	0.383300	0.292300
1.106100	0.949200	0.605500	0.416900	0.317800
1.288000	1.105200	0.705100	0.485400	0.370100
1.562800	1.341000	0.855500	0.589000	0.449100
1.930400	1.656500	1.056800	0.727500	0.554700

Table 10.18 Load of ψ and C*.

M = 8, $\Phi_0+\Phi_1= 2$, σ* = 0.05, α* = −0.05, ε*=−0.05, β*=0.1

ψ→
C* ↓

3.179000	2.727900	1.740300	1.198100	0.913500
2.495200	2.141200	1.366000	0.940400	0.717000
1.956500	1.678900	1.071100	0.737400	0.562200
1.562800	1.341000	0.855500	0.589000	0.449100
1.314000	1.127600	0.719300	0.495200	0.377600

Table 10.19 Load of σ* and α*.

M = 8, $\Phi_0+\Phi_1= 2$, ψ = 0.01, β* = 0.1, ε*=−0.05   *C*=0.1*

σ*→
α* ↓

1.608600	1.592800	1.547100	1.476500	1.387900
1.352900	1.341000	1.306700	1.253300	1.185500
1.149800	1.140700	1.114600	1.073500	1.020800
0.986100	0.979200	0.958900	0.926800	0.885400
0.852700	0.847200	0.831300	0.806000	0.773000

Table 10.20 Load of σ and ε*.

M = 8, $\Phi_0+\Phi_1= 2$, ψ = 0.01, β* = 0.1, α*=−0.05   *C*=0.1*

σ*→
ε* ↓

1.442200	1.428700	1.389900	1.329600	1.253500
1.352900	1.341000	1.306700	1.253300	1.185500
1.274000	1.263500	1.233000	1.185300	1.124500
1.203800	1.194400	1.167100	1.124300	1.069400
1.140900	1.132500	1.107900	1.069300	1.019500

Table 10.21 Load of σ* and B*.

M = 8, $\Phi_0+\Phi_1= 2$, ψ = 0.01, ε* = −0.05, α*=−0.05   *C*=0.1*

σ*→
β* ↓

0.881200	0.872800	0.848900	0.812600	0.768000
0.956700	0.949200	0.927300	0.893400	0.850400
1.113900	1.105200	1.079900	1.040300	0.989500
1.352900	1.341000	1.306700	1.253300	1.185500
1.673500	1.656500	1.607700	1.532500	1.438200

Table 10.22 Load of σ^* and C^*.

M = 8, $\Phi_0+\Phi_1= 2$, $\psi = 0.01$, $\varepsilon^* = -0.05$, $\alpha^*=-0.05$, $\beta^*= 0.1$				
$\sigma^*\rightarrow$				
$C^*\downarrow$ 2.764000	2.727900	2.625000	2.468800	2.277700
2.166500	2.141200	2.068500	1.957600	1.820500
1.696200	1.678900	1.629100	1.552400	1.456400
1.352900	1.341000	1.306700	1.253300	1.185500
1.136600	1.127600	1.101300	1.060200	1.007700

Table 10.23 Load of α^* and ε^*.

M = 8, $\Phi_0+\Phi_1= 2$, $\psi = 0.01$, $\beta^* = 0.1$, $\sigma^*=0.05$				$C^*=0.1$
$\alpha^*\rightarrow$				
$\varepsilon^*\downarrow$ 1.718100	1.428700	1.203600	1.025100	0.881400
1.592800	1.341000	1.140700	0.979200	0.847200
1.484600	1.263500	1.084100	0.937200	0.815600
1.390100	1.194400	1.032900	0.898600	0.786200
1.306900	1.132500	0.986200	0.863100	0.758900

Table 10.24 Load of α^* and B^*.

M = 8, $\Phi_0+\Phi_1= 2$, $\psi = 0.01$, $\varepsilon^* = -0.05$, $\sigma^*=0.05$				$C^*=0.1$
$\alpha^*\rightarrow$				
$\beta^*\downarrow$ 1.041400	0.872800	0.742900	0.640100	0.557100
1.115000	0.949200	0.816600	0.708800	0.620000
1.298800	1.105200	0.949200	0.821900	0.716700
1.592800	1.341000	1.140700	0.979200	0.847200
1.997100	1.656500	1.391100	1.180800	1.011500

Table 10.25 Load of α^* and C^*.

M = 8, $\Phi_0+\Phi_1= 2$, $\psi = 0.01$, $\varepsilon^* = -0.05$, $\sigma^*=0.05$, $\beta^*=0.1$				
$\alpha^*\rightarrow$				
$C^*\downarrow$ 3.391000	2.727900	2.229900	1.847700	1.549100
2.624800	2.141200	1.772100	1.484800	1.708900
2.025400	1.678900	1.409100	1.195300	1.023500
1.592800	1.341000	1.140700	0.979200	0.847200
1.327100	1.127600	0.967200	0.836400	0.728700

Table 10.26 Load of ε* and B*.

M = 8, $\Phi_0+\Phi_1$= 2, ψ = 0.01, α* = −0.05, σ*=0.05				C*=0.1

ε*→

β*↓				
0.947100	0.872800	0.813400	0.764800	0.724200
1.006100	0.949200	0.900300	0.857700	0.820000
1.166600	1.105200	1.050300	1.000900	0.956100
1.428700	1.341000	1.263500	1.194400	1.132500
1.792400	1.656500	1.539700	1.438200	1.349200

Table 10.27 Load of ε* and C*.

M = 8, $\Phi_0+\Phi_1$= 2, ψ = 0.01, α* = −0.05, σ*=0.05,β*=0.1

ε*→

C* ↓				
3.052800	2.727900	2.460200	2.236400	2.047200
2.359100	2.141200	1.958500	1.803300	1.670100
1.817800	1.678900	1.559600	1.455900	1.365200
1.428700	1.341000	1.263500	1.194400	1.132500
1.192000	1.127600	1.070200	1.018600	0.972000

Table 10.28 Load of B* and C*.

M = 8, $\Phi_0+\Phi_1$= 2, ψ = 0.01, ε* = −0.05, σ*=0.05, α* = −0.05

β*

C↓				
1.341000	1.723600	2.185900	2.727900	3.349700
1.060500	1.341000	1.701200	2.141200	2.660900
0.904500	1.082900	1.341000	1.678900	2.096500
0.872800	0.949200	1.105200	1.341000	1.656500
0.965600	0.939900	0.993900	1.127600	1.341000

Justifications

- When the roughness parameters are taken to be zero, the analysis of Prajapati (1995) becomes the basis for this study. The results for hydromagnetic squeeze films among two conducting uneven surfaces are obtained for non-spongy conducting plates. The study tends to squeeze non-magnetic spongy films between rough spongy plates in the limiting situation of M→0.
- This analysis basically addresses the study of the behavior of a hydro-magnetic squeeze film among rough spongy plates when ϕ_1 and ϕ_0 are taken to be nil.

Conclusion

It is evident that as plate conductivities, the curvature parameter for the top plate, and negative variance rise, the lubricant pressure, Both the response time and the load bearing capability grow. Additionally, even if M and ϕ_0 and ϕ_1 have been properly chosen and taken into account, this article makes it evident that roughness parameters and curvature parameters must be given full care when building the bearing system.

Conflict of Interest: There is no competing interest amongst the authors.

Funding: The research, writing, and/or publication of this article were not supported financially by the author(s).

Acknowledgment

The reviewers' and editor's criticisms and recommendations are gratefully accepted by the authors.

References

1.   Andharia, P. I., Gupta, J. L., and Deheri, G. M. (1997). Effect of longitudinal surface roughness on hydrodynamic lubrication of slider bearings. *Proc. Tenth Int. Conf. Surf. Modif. Technol.*, 872–880.
2.   Andharia, P. I., Gupta, J. L., and Deheri, G. M. (1999). Effect of transverse surface roughness on the behaviour of squeeze film in a spherical bearing. *J. Appl. Mech. Engg.*, 4, 19–24.
3.   Berthe, D. and Godet, M. (1973). A more general form of Reynolds equation–Application to rough surfaces. *Wear*, 27, 345–357.
4.   Burton, R. A. (1963). Effect of two-dimensional sinusoidal roughness on the load support characteristics of a lubricant film. *J. Basic Engg. Trans. ASME*, 85, 258–264.
5.   Christensen, H. and Tonder, K. C. (1969 a). Tribology of rough surfaces: Stochastic models of hydrodynamic lubrication. *SINTEF Report No. 10*, 69–18.
6.   Christensen, H. and Tonder, K. C. (1969 b). Tribology of rough surfaces: Parametric study and comparison of lubrication models. *SINTEF Report No. 22*, 69–18.
7.   Christensen, H. and Tonder, K. C. (1970). Tribology of rough surfaces: A stochastic model of mixed lubrication. *SINTEF Report No. 18*, 70–21.
8.   Davis, M. G. (1963). The Generation of pressure between rough lubricated, moving deformable surfaces. *Lub. Engg.*, 19, 246.
9.   Dodge, F. T., Osterle, J. F., and Rouleau, W. T. (1965). Magnetohydrodynamic squeeze film bearings. *J. Basic Engg. Trans. ASME*, 87, 805–809.
10. Elco, R. A. and Huges, W. F. (1962). Magnetohydrodynamic pressurization in liquid metal lubrication. *Wear*, 5, 198–207.
11. Guha, S. K. (1993). Analysis of dynamic characteristics of hydrodynamic journal bearings with isotropic roughness effects. *Wear*, 167, 173–179.
12. Gupta, J. L. and Deheri, G. M. (1996). Effect of roughness on the behaviour of squeeze film in a spherical bearing. *Tribol. Trans.*, 39, 99–102.
13. Kuzma, D. C. (1964). Magnetohydrodynamic squeeze films. *J. Basic Engg. Trans. ASME*, 86, 441–444.
14. Kuzma, D. C., Maki, E. R., and Donnelly, R. J. (1964). The magnetohydrodynamic squeeze films. *J. Fluid Mech.*, 19, 395–400.
15. Maki, E. R., Kuzma, D. C., and Donnelly, R. J. (1966). Magnetohydrodynamic lubrication flow between parallel plates. *J. Fluid Mech.*, 26(3), 537–543.

16. Michell, A. G. M. (1950). Lubrication, its principle and practice. London: Blackie. p. 317.
17. Mishra, S. R., Barik, M., and Dash, G. C. (2018). An analysis of hydrodynamic ferro-fluid lubrication of an inclined rough slider bearing. *Tribol. Mat. Surf. Interf.*, 12(1), 17–26.
18. Vadher, P. A., Vinodkumar, P. C., Deheri, G. M., and Patel, R. M. (2008). Behavior of hydromagnetic squeeze films between two conducting rough porous circular plates. *J. Engg. Tribol.*, 222(4), 569–579.
19. Patel, K. C. (1975). Hydromagnetic squeeze film with slip velocity between two porous annular disks. *J. Lub. Technol. Trans. ASME*, 97, 644–647.
20. Patel, K. C. and Gupta, J. L. (1979). Behaviour of hydromagnetic squeeze film between porous plates. *Wear*, 56, 327–339.
21. Patel, K. C. and Hingu, J. V. (1978). Hydromagnetic squeeze film behaviour in porous circular disks. *Wear*, 49, 239–246.
22. Patel, R. M. and Deheri, G. M. (2003). Magnetic fluid-based squeeze film behavior between rotating porous circular plates with a concentric circular pocket and surface roughness effects. *Int. J. Appl. Mech. Engg.*, 8(2), 271–277.
23. Patel, R. M. and Deheri, G. M. (2004). Magnetic fluid-based squeeze film behaviour between annular plates and surface roughness effect. *AIMETA, Int. Tribol. Conf.*, 631–638.
24. Prajapati, B. L. (1991). Behaviour of squeeze film between rotating porous circular plates: Surface roughness and elastic deformation effects. *J Pure Appl. Math. Sci.*, 33(1–2), 27–36.
25. Prajapati, B. L. (1992). Squeeze film behaviour between rotating porous circular plates with a concentric circular pocket: Surface roughness and elastic deformation effects. *Wear*, 152, 301–307.
26. Prajapati, B. L. (1995). On certain theoretical studies in hydrodynamic and electro-magnetohydrodynamic lubrication. Ph.D Thesis, S. P. University, Vallabh Vidyanagar, (Gujarat), India.
27. Prakash, J. and Tiwari, K. (1982). Lubrication of a porous bearing with surface corrugations. *J Lub. Technol. Trans. ASME*, 104, 127–134.
28. Prakash, J. and Vij, S. K. (1973). Hydrodynamic lubrication of a porous slider. *J. Mech. Engg. Sci.*, 15, 232–234.
29. Shukla, J. B. (1965). Hydromagnetic theory of squeeze films. *ASME*, 87, 142.
30. Shukla, J. B. and Prasad, R. (1965). Hydromagnetic squeeze films between two conducting surfaces. *J. Basic Engg. Trans. ASME*, 87, 818–822.
31. Sinha, P. C. and Gupta, J. L. (1973). Hydromagnetic squeeze films between porous rectangular plates. *J. Lub. Technol. Trans. ASME*, F95, 394–398.
32. Sinha, P. C. and Gupta, J. L. (1974). Hydromagnetic squeeze films between porous annular disks. *J. Math. Phy. Sci.*, 8, 413–422.
33. Snyder, W. T. (1962). The magnetohydrodynamic slider bearings. *J. Basic Engg. Trans. ASME*, 85, 429–434.
34. Ting, L. L. (1975). Engagement behaviour of lubricated porous annular disks Part I: Squeeze film phase, surface roughness and elastic deformation effects. *Wear*, 34, 159–182.
35. Tonder, K. C. (1967). Surface distributed waviness and roughness. *First World Conference in Industrial Tribology*, New Delhi, 1972, A3, 128.
36. Tonder, K. C. (1977). Lubrication of surfaces having area distributed isotropic roughness. *J. Lub. Tech. Trans. ASME*, 99, 323–330.
37. Tzeng, S. T. and Saibel, E. (1967). Surface roughness effect on slider bearing lubrication. *J. Lub. Technol. Trans. ASME*, 10, 334–338.

11 Performance of longitudinally rough circular step bearings with hydromagnetic squeeze films considering the effect of deformation and slip velocity

Dr. Jatinkumar V. Adeshara[1], Dr. Suresh G. Sorathiya[2], Dr. Hardik P. Patel[3], Dr. Gunamani B. Deheri[4] and Rakesh M. Patel[5]

[1]Assistant Professor, Vishwakarma Government Engineering College, Chandkheda, Ahmedabad-382424, Gujarat, India

[2]Assistant Professor, Department of Science, Sal Institute of Technology & Engineering Research, Ahmedabad, India

[3]Assistant Professor, Department of Humanity and Science, L. J. Institute of Engineering and Technology, LJK University, Ahmedabad, India

[4]Retired Associate Professor, Department of Mathematics, S. P. University, Vallabh Vidyanagar-388120, Gujarat, India

[5]Assistant Professor Department of Mathematics, Gujarat Arts and Science College, Ahmedabad-380006, Gujarat, India

Abstract

Objectives: Investigating the collective effects of slip velocity and deformation on spongy circular step bearings on hydro magnetic squeeze film with suitable B.C. is the purpose of the study. **Methods/statistical analysis:** The stochastic Reynolds' equation for the longitudinal roughness pattern is developed by Christensen and Tonder using their stochastic model. Also the outcome of irregularity parameters numerically modeled. The resulting governing equation is solved by analytically method. In order to determine the pressure distribution; the corresponding stochastically averaged Reynolds equation must be solved. The outcomes obtained here are represented by a graph. **Findings:** As a result, the load carrying capacity is calculated. All of the graphical data show that the longitudinal roughness, when combined with the deformation, has a significant detrimental impact on the bearing system's performance. The superiority of the longitudinal surface roughness effects over the transverse surface roughness has been demonstrated beyond any reasonable doubt. However, the current work shows that in the situation of negatively skewed roughness, the detrimental effect of porosity can be somewhat offset by the favorable result of magnetization parameters by appropriately selecting the aspect ratio. Furthermore, when negative variance is present, this compensation is even more obvious.

Keywords: Reynolds' equation, roughness, squeeze film, hydromagnetic fluid

Introduction

When analyzing the behavior of the S.F. between spongy annular discs, Wu (1971) took into consideration. The S.F. performance in spongy circular discs was

——◆—— $\phi 0+\phi 1=0.5$ ——■—— $\phi 0+\phi 1=1.5$ ——▲—— $\phi 0+\phi 1=2.5$

——✳—— $\phi 0+\phi 1=3.5$ ——✴—— $\phi 0+\phi 1=4.5$

——◆—— $\varepsilon^*=-0.10$ ——■—— $\varepsilon^*=-0.05$ ——▲—— $\varepsilon^*=0$

——✳—— $\varepsilon^*=0.05$ ——✴—— $\varepsilon^*=0.10$

——◆—— $k=0.35$ ——■—— $k=0.45$ ——▲—— $k=0.55$ ——✳—— $k=0.65$ ——✴—— $k=0.75$

——◆—— $\alpha^*=-0.10$ ——■—— $\alpha^*=-0.05$ ——▲—— $\alpha^*=0$

——✳—— $\alpha^*=0.05$ ——✴—— $\alpha^*=0.10$

——◆—— $s^*=0$ ——■—— $s^*=0.05$ ——▲—— $s^*=0.1$ ——✳—— $s^*=0.15$ ——✴—— $s^*=0.2$

——◆—— $\psi=0.001$ ——■—— $\psi=0.003$ ——▲—— $\psi=0.005$

——✳—— $\psi=0.007$ ——✴—— $\psi=0.009$

——◆—— $p^*=0$ ——■—— $p^*=0.0001$ ——▲—— $p^*=0.001$

——✳—— $p^*=0.01$ ——✴—— $p^*=0.1$

addressed by Murti (1974). The M.C. approximation was utilized by Prakash and Vij (1973) to analyses. S.F. problems involving diverse shapes. The S.F. between rotating spongy annular discs was examined by Wu (1971).

Ting (1975) looked into the issue of spongy clutch plates engaging, employing annular plates to simulate the porous housing's surface irregularities and the effects of deformation. However, it is generally known that due to logarithmic singularities, this method cannot yield the analogous analysis for spongy circular plates. Along with the impacts of elastic deformations and the irregularities of the contact surfaces, mechanical elements are frequently used in practical applications, additionally, pockets might appear as dents and cavities.

These pockets which are frequently circular and concentric in shape are caused by the material's deterioration from rotational motion in some situations. Therefore, it is crucial to research the lubrication of spongy metal in such machine components. For the first time, Archibald (1956) gave an analysis of the S.F. that forms between circular, rigid, smooth, non-porous plates with concentric pockets. The studies mentioned above all focused on traditional lubricants. With the external magnetic field oblique to the outer disc. Bhat and Deheri (1991) used a magnetic lubricant to analyze the S.F. behavior between spongy annular discs. It was determined that the use of magnetic fluid as a lubricant improved performance traits generally.

Additionally, it was discovered that the M.F. lubricant performed far better than the S.D. Bhat and Deheri (1993) conducted additional research on how the S.F. behaves in curved, spongy circular plates when an M.F. lubricant is present. The study given was expanded by Shah and Bhat (2000) by taking rotation into account. However, the effects of uneven wear and elastic thermal, real-world configurations are typically anything but smooth. After some wear and run-in, the bearing surfaces also become rough. Roughness has a seemingly random aspect and doesn't seem to adhere to any particular structural structure.

The analysis of Tzeng and Saibel (1967) was modified and developed by C.&T., Christensen and Tonder (1969a), Christensen and Tonder (1970) and Tzeng and Saibel (1967), who also provided a thorough general analysis for both T.R. and L.R. This approach by C.&T., Christensen and Tonder (1970) and Tzeng and Saibel (1967) and Prakash and Tiwari (1982) afterwards served as the foundation for numerous investigations looking at the impact of surface irregularity (Ting, 1975; Prakash and Tiwari, 1982; Prajapati, 1992; Guha, 1993; Gupta and Deheri, 1996). In their examination into the behavior of a squeezing film made of magnetic fluid between annular plates. Patel and Deheri (2004) also looked at the impact of T.R., Prajapati (1991) discussed the S.F. behavior among rotating spongy circular plates with a concentric pocket while only taking into account a specific kind of irregularity.

Analysis

The geometry and configuration of the bearing system is presented below.

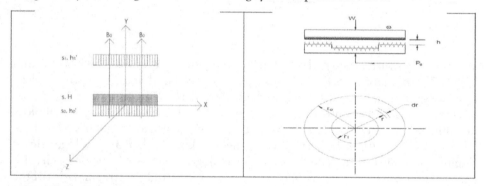

As seen in the above Figure, a thrust load is applied, and the bearing carries the load without coming into touch with any metal. The fluid in the pocket and on the surface carries the load w. The fluid leaves the recess in a radial direction due to land or sill limits. Rough surfaces are present (longitudinally). From C.&T. (Tzeng and Saibel, 1967; Christensen and Tonder, 1970; Prakash and Tiwari, 1982), the film thickness is brought.

The pressure-induced flow for a circular step bearing is represented by the Reynolds' type equation in concern as (Majumdar, 1986; Patel, Deheri and Vadher, 2004).

$$Q = -\frac{2\pi r \frac{dp}{dr}\left[\frac{2}{M^3 A}\left[\frac{M}{2}-\tanh\frac{M}{2}\right]+\frac{\psi}{c^2 A}\right]\left[\frac{\varphi_0+\varphi_1+1}{\varphi_0+\varphi_1+\left(\tanh\frac{M}{2}\right)/\left(\frac{M}{2}\right)}\right]}{12\mu} \qquad (1)$$

$$A = (h + p_a p'\delta)^{-3}\left[\frac{1+2sh}{1+4sh}\right]^{-3}\left[1 - 3\alpha(h + p_a p'\delta)^{-1}\left[\frac{1+2sh}{1+4sh}\right]^{-1/3} - 3(h + p_a p'\delta)^{-2}\left[\frac{1+2sh}{1+4sh}\right]^{-2/3} -\frac{3}{40}(h + p_a p'\delta)^{-3}(3\sigma^2\alpha + \alpha^3 + \varepsilon)\left[\frac{1+2sh}{1+4sh}\right]^{-1}\right]$$

Reynolds' boundary conditions are used as

$$r = r_o; \quad p = 0; \quad r = r_i; \quad p = ps; \tag{2}$$

(2) The formula for the film pressure's governing equation is

$$p = p_s \frac{\ln\left(\frac{r}{r_0}\right)}{\ln\left(\frac{r_i}{r_0}\right)}$$

$$p_s = \frac{6}{\pi\left[\frac{2}{M^3 A}\left[\frac{M}{2} - \tanh\frac{M}{2}\right] + \frac{\psi}{c^2 A}\right]\left[\frac{\varphi_0 + \varphi_1 + 1}{\varphi_0 + \varphi_1 + \left(\tanh\frac{M}{2}\right)/\left(\frac{M}{2}\right)}\right]} \ln\left(\frac{r_o}{r_i}\right)$$

Introducing the non-dimensional quantities

$$P_s *= \frac{6\ln\left(\frac{1}{k}\right)}{\pi\left[\frac{2}{M^3 B}\left[\frac{M}{2} - \tanh\frac{M}{2}\right] + \frac{\bar{\psi}}{Bc^2}\right]\left[\frac{\varphi_0 + \varphi_1 + 1}{\varphi_0 + \varphi_1 + \left(\tanh\frac{M}{2}\right)/\left(\frac{M}{2}\right)}\right]} \tag{3}$$

$$B = H^{-3}(1 + \bar{p}\delta)^{-3}\left[\frac{1 + 2\bar{s}}{1 + 4\bar{s}}\right]^{-3} * [1 - 3\bar{\alpha}(1 + \bar{p}\delta)^{-1}\left[\frac{1 + 2\bar{s}}{1 + 4\bar{s}}\right]^{-\frac{1}{3}}$$

$$- 3(1 + \bar{p}\delta)^{-2}(\bar{\sigma}^2 + \bar{\alpha}^2)\left[\frac{1 + 2\bar{s}}{1 + 4\bar{s}}\right]^{-\frac{2}{3}} - \frac{3}{40}(1 + \bar{p}\delta)^{-3}(3\bar{\sigma}^2\bar{\alpha} + \bar{\alpha}^3 + \bar{\varepsilon})\left[\frac{1 + 2\bar{s}}{1 + 4\bar{s}}\right]^{-1}$$

$$\psi = \frac{\varphi H_0}{h^3}, \qquad \bar{\sigma} = \frac{\sigma}{h}, \qquad \bar{\alpha} = \frac{\alpha}{h}, \qquad \bar{\varepsilon} = \frac{\varepsilon}{h^3},$$

$$\bar{p} = p_a p', \qquad \bar{\delta} = \frac{\delta}{h'}, \qquad \bar{s} = sh$$

allows for the N.D. description of P.D. as

$$P = P_s * \frac{\ln\left(\frac{r}{r_0}\right)}{\ln\left(\frac{r_i}{r_0}\right)} \tag{4}$$

By integrating the pressure which has a dimensionless shape, the L.B.C. w is determined.

$$W = \frac{P_s^* \ln(1 - k^2)}{\ln\left(\frac{1}{k}\right)} \tag{5}$$

Results and discussions

It is easily seen that the non-dimensional pressure distribution is given by equation (3), while the dimensionless L.C.C. is determined from equation (4). Depending on the distribution of pressure and load, the impact of conductivity parameters is calculated by

$$\frac{\varphi_0 + \varphi_1 + \left(\tanh\frac{M}{2}\right)/\left(\frac{M}{2}\right)}{\varphi_0 + \varphi_1 + 1}$$

This turns to

$$\frac{\varphi_0 + \varphi_1}{\varphi_0 + \varphi_1 + 1}$$

Due to the fact that $(2/M) \cong 2$ and $\tanh(M) \cong 1$ for enormous values of M. Additionally, it is noted that as $\phi 0 + \phi 1$ increases, the pressure and load also do so because both functions are growing functions of $\phi 0 + \phi 1$.

For several values of $\phi 0 + \phi 1, \sigma^*, \alpha^*, \varepsilon^*, k, s^*$, several values of $\psi^* \delta^*$ and p^*, respectively the fluctuation of L.C.C. with magnetization parameter is illustrated in Figures 11.1–11.9. It is evident that the M.F. lubricant considerably boosts the L.T.C. As shown in Figures 11.10–11.17, conductivity has the impact of increasing load.

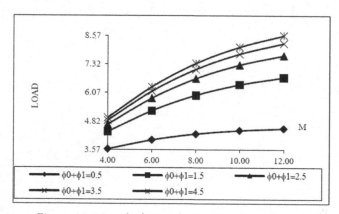

Figure 11.1 Load of M and $\phi 0 + \phi 1$

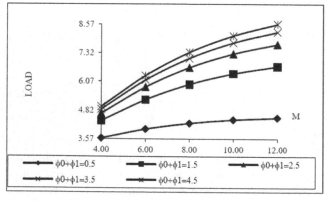

Figure 11.2 Load of M and σ^*

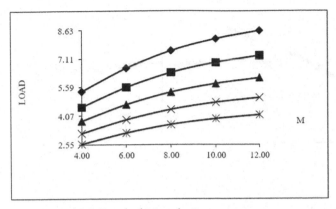

Figure 11.3 Load of M and α*

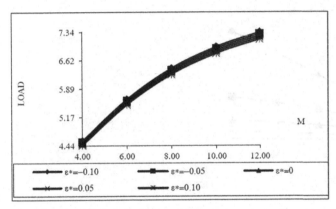

Figure 11.4 Load of M and ε*

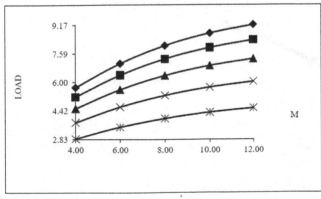

Figure 11.5 Load of M and k

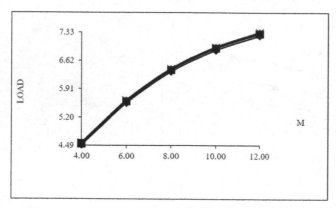

Figure 11.6 Load of M and s*

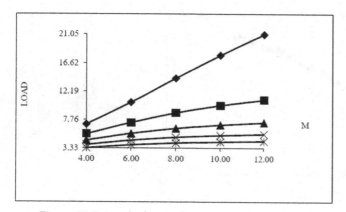

Figure 11.7 Load of M and ψ*

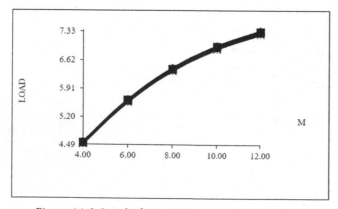

Figure 11.8 Load of M and δ*

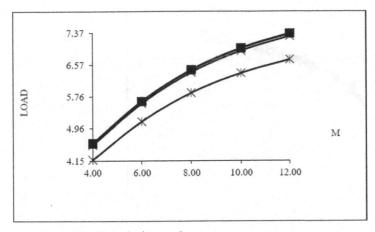

Figure 11.9 Load of M and p

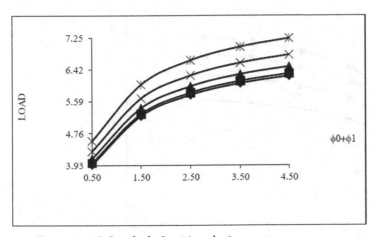

Figure 11.10 Load of $\phi 0 + \phi 1$ and σ^*

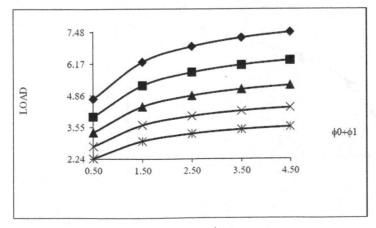

Figure 11.11 Load of $\phi 0 + \phi 1$ and α^*

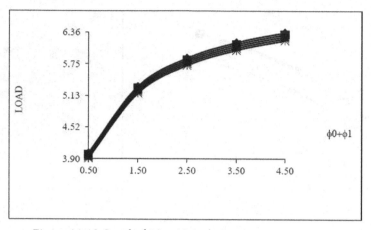

Figure 11.12 Load of $\phi 0 + \phi 1$ and ε^*

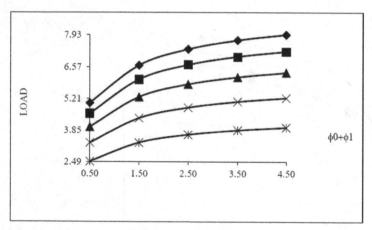

Figure 11.13 Load of $\phi 0 + \phi 1$ and k

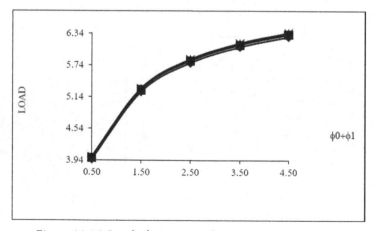

Figure 11.14 Load of $\phi 0 + \phi 1$ and s*

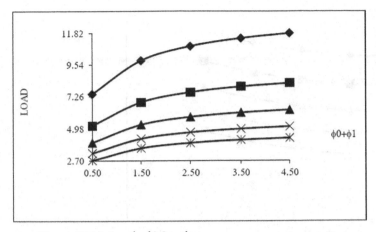

Figure 11.15 Load of M and ψ

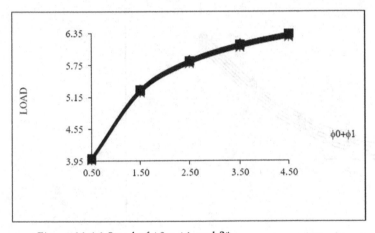

Figure 11.16 Load of φ0 + φ1 and δ*

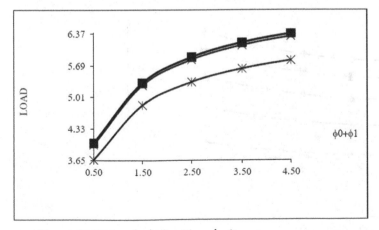

Figure 11.17 Load of φ0 + φ1 and p*

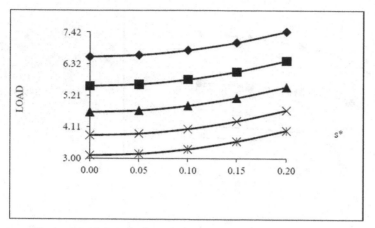

Figure 11.18 Load of σ* and α*

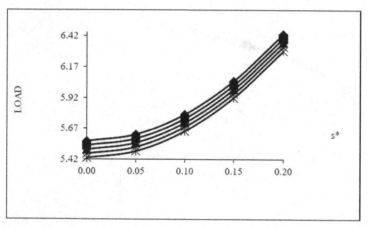

Figure 11.19 Load of σ* and ε*

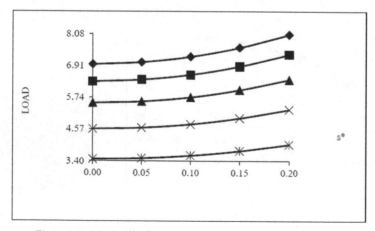

Figure 11.20 Load of σ* and k

Figures 11.18–11.24 show how the distribution of load is affected by the S.D. associated with irregularity. It is obvious that the S.D. has a positive impact on the bearing system because it greatly increases the load.

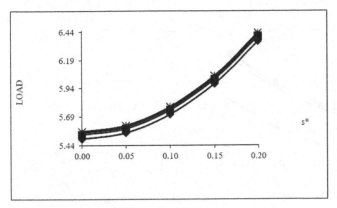

Figure 11.21 Load of σ* and s*

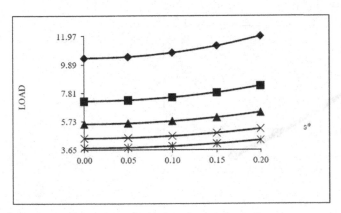

Figure 11.22 Load of σ* and ψ*

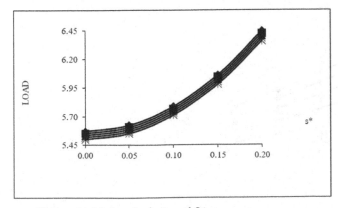

Figure 11.23 Load of σ* and δ*

The profile for the fluctuation of load with respect to variance is depicted in Figures 11.25–11.30. It is observed that the load reduces owing to variance (+ve), whereas the L.C.C. increases as a result of variance (-ve). It is discovered that lower values of cause a not so significant effect while larger values of cause a minimal effect.

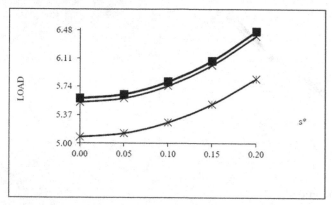

Figure 11.24 Load of α^* and p^*

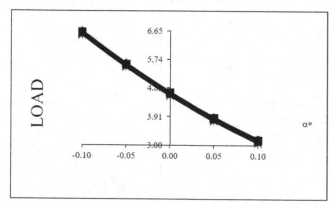

Figure 11.25 Load of α^* and ε^*

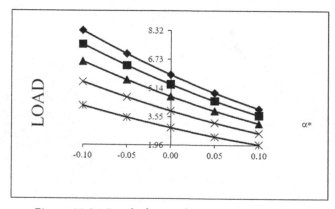

Figure 11.26 Load of α^* and k

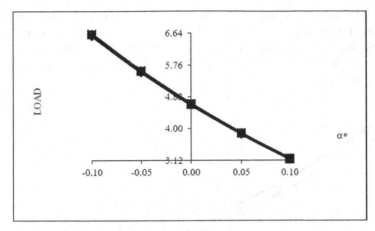

Figure 11.27 Load of α* and s*

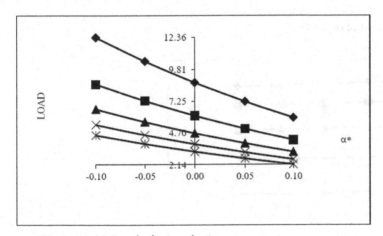

Figure 11.28 Load of α* and ψ*

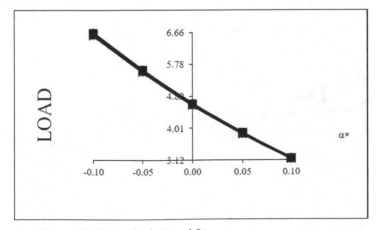

Figure 11.29 Load of α* and δ*

Figures 11.31–11.35 show how skewness affects the distribution of load. The skewness is found to follow the trends of α*. These demonstrate that the combined effect of variation (-ve) and negatively skewed irregularity is significantly

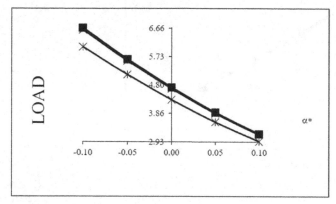

Figure 11.30 Load of α* and p*

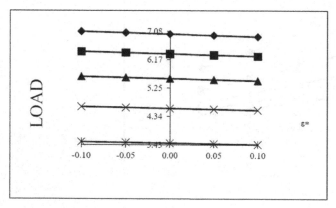

Figure 11.31 Load of ε* and k

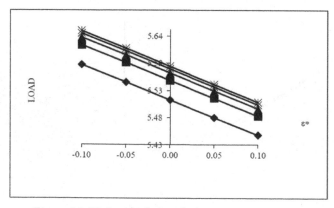

Figure 11.32 Load of ε* and s*

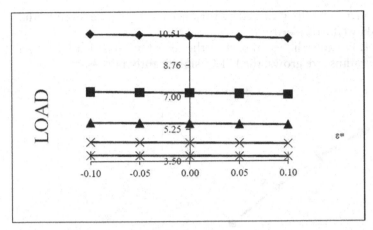

Figure 11.33 Load of ε* and ψ

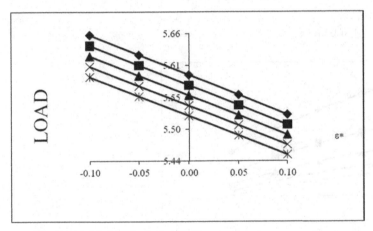

Figure 11.34 Load of ε* and δ*

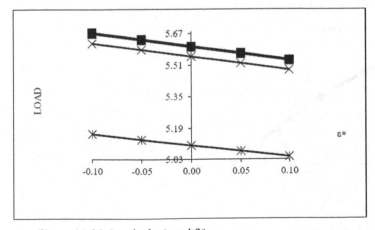

Figure 11.35 Load of α* and δ*

beneficial since the increase in load caused by variance (-ve) increases even further due to negatively skewed irregularity.

Figures 11.36–11.39 shows how radii affect the distribution of load. It is evident that when the radius size grows, the L.T.C. significantly falls.

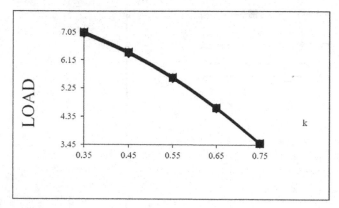

Figure 11.36 Load of k and s

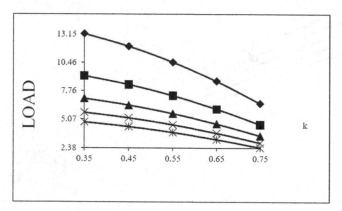

Figure 11.37 Load of k and ε

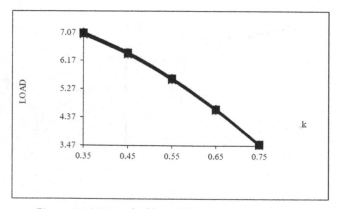

Figure 11.38 Load of k and δ*

Figures 11.40–11.42 shows that the load of slip parameter increases with respect to porosity, deformation and ambient pressure. Whereas Figures 11.43 and 11.44 shows the effect of porosity with respect to deformation and ambient

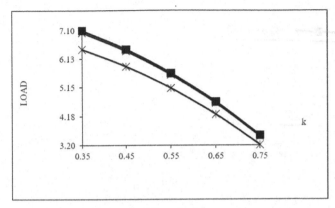

Figure 11.39 Load of k and p*

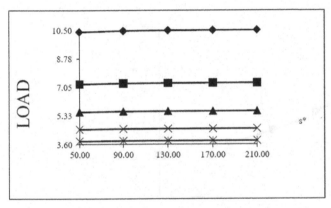

Figure 11.40 Load of s* and ψ

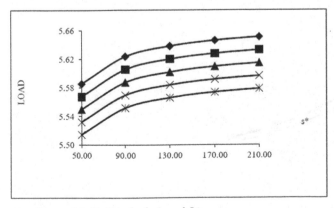

Figure 11.41 Load of s* and δ*

pressure which shows that the higher value of deformation and ambient pressure decreases the load. Figure 11.45 conclude the load decreases, deformation with respect to pressure p*.

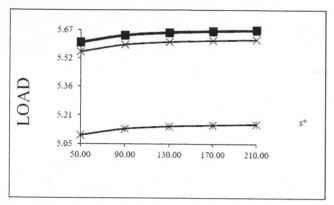

Figure 11.42 Load of s* and p*

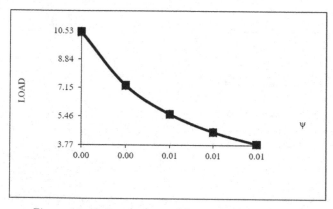

Figure 11.43 Load of ψ and δ*

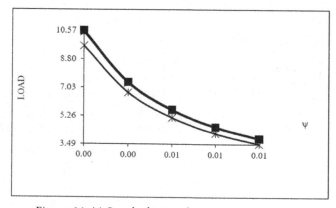

Figure 11.44 Load of ψ* and p*

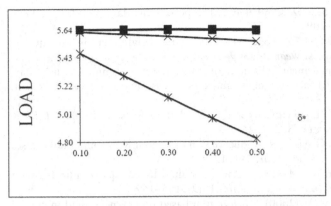

Figure 11.45 Load of δ* and p*

Conclusion

- This investigation clears that although the combined effect of deformation and slip is considered adverse; hydro magnetic retrieves the situation to some extent in the case of longitudinally rough surface.
- However, when the deformation and slip are at lowest level the combine effect of longitudinally roughness and hydro magnetic is quite significant.
- The graphical representation suggests that by suitably chosen the parameters, this study could be useful in industries aspect.

Abbreviations

T.R.   Transverse Roughness
L.R.   Longitudinal Roughness
L.C.C. Load Carrying Capacity
L.B.C. Load Bearing Capacity
L.T.C. Load Taking Capacity
S.F.   Squeeze Film
M.F.   Magnetic Field
S.D.   Standard Deviation
M.C.   Morgan–Cameron
C.&T.   Christensen and Tonder's
P.D.   Pressure Distribution
B. C.   Boundary Condition

Acknowledgment

The authors acknowledge with thanks the comments and suggestions of the reviewers/Editor.

References

1.   Wu, H. (1971). The squeeze film between rotating porous annular disks. *Wear*, 18, 461–470.

2.   Murti, P. R. K. (1974). Squeeze film behaviour in porous circular disks. *J. Lub. Tech. Trans. ASME,* 96, 206–209.

3.   Prakash, J. and Vij, S. K. (1973). Load capacity and time height relations for squeeze film between porous plates. *Wear,* 24, 309–322.

4.   Ting, L. L. (1975). Engagement behaviour of lubricated porous annular disks Part I: Squeeze film phase, surface roughness and elastic deformation effects. *Wear,* 34, 159–182.

5.   Archibald, F. R. (1956). Load capacity and time relations for squeeze films. *J. Basic Engg. Trans. ASME, Sear. D* 78, 231–245.

6.   Bhat, M. V. and Deheri, G. M. (1991). Squeeze film behaviour in porous annular discs lubricated with magnetic fluid. *Wear,* 151, 123.

7.   Bhat, M. V. and Deheri, G. M. (1993). Magnetic fluid based squeeze film between porous circular disks. *J. Indian Acad. Math.,* 15(2), 145–147.

8.   Shah, R. C. and Bhat, M. V. (2000). Squeeze film based on magnetic fluid in curved porous rotating circular plates. *J. Magnet. Magnet. Mat.,* 208(1), 115–119.

9.   Christensen, H. and Tonder, K. C. (1969a). Tribology of rough surfaces: stochastic models of hydrodynamic lubrication. SINTEF Report No.10/69-18. doi: 10.12691/ijml-2-1-7.

10.   Christensen, H. and Tonder, K. C. (1969b). Tribology of rough surfaces: parametric study and comparison of lubrication model. SINTEF Report No.22/69-18. doi: 10.12691/ijml-2-1-7.

11.   Christensen, H. and Tonder, K. C. (1970). The hydrodynamic lubrication of rough bearing surfaces of finite width. *ASME-ASLE Lub. Conf.,* Paper no.70-lub-7.

12.   Tzeng, S. T. and Saibel, E. (1967). Surface roughness effect on slider bearing lubrication. *ASLE Trans.,* 10, 334–338. doi.org/10.1080/05698196708972191.

13.   Prakash, J. and Tiwari, K. (1982), Lubrication of a porous bearing with surface corrugations. *J. Lub. Technol. Trans. ASME,* 104, 127–134.

14.   Prajapati, B. L. (1991). Behavior of squeeze film between rotating porous circular plates: Surface roughness and elastic deformation effects. *J. Pure Appl. Math. Sci.,* 33(1–2), 27–36.

15.   Prajapati, B. L. (1992). Squeeze film behavior between rotating porous circular plates with a concentric circular pocket: Surface roughness and elastic deformation effects. *Wear,* 152, 301–307.

16.   Guha, S. K. (1993). Analysis of dynamic characteristics of hydrodynamic journal bearings with isotropic roughness effects. *Wear,* 167, 173–179.

17.   Gupta, J. L. and Deheri, G. M. (1996). Effect of roughness on the behaviour of squeeze film in a spherical bearing. *Tribol. Trans.,* 39, 99–102.

18.   Patel, R. M. and Deheri, G. M. (2004). Magnetic fluid based squeeze film behavior between annular plates and surface roughness effect. *AIMETA, Int. Tribol. Conf. Rome, Italy,* 631–638.

19.   Majumdar, B. C. (1986). Introduction to tribology of bearings. A. H. Wheeler and Co.

12 Bearing deformation and longitudinal roughness effects on the performance of a hydro magnetic-based squeeze film with velocity slip on truncated conical plates

Dr. H. P. Patel[1], Dr. J. V. Adeshara[2], Dr. M. P. Shekhawat[3], Dr. G. M. Deheri[4] and R. M. Patel[4,a]

[1]Department of Humanity and Science, LJIET, L J K University. Ahmedabad, Gujarat, India

[2]Assistant Professor, Vishwakarma Government Engineering College, Chandkheda, Ahmedabad-382424, Gujarat, India

[3]Department of Mathematics, Gujarat Arts and Science College, Ahmedabad-380006, Gujarat, India

[4]Department of Mathematics, S. P. University, Vallabh Vidyanagar-388120, Gujarat, India

Abstract

This work describes the bearing deformation and slip velocity that occur as a result of the interaction in hydromagnetic squeeze film and longitudinal irregularity factor on truncated rough conical plate. During the investigation, the hydromagnetic lubrication theory and the slip model developed by Beavers and Joseph were used. Additionally, the generalized Reynolds type equations for the pressure of the fluid film were obtained by applying the appropriate boundary conditions. Finally, the expressions for distribution of pressure and bearing's load capacity as a function of slip parameter, roughness parameter, and Hartmann number were derived. The stochastic averaging model that Christensen and Tonder developed has been utilized in this study so that the influence of longitudinal roughness can be determined. It has come to light that the implementation of the magnetic field effect results in a discernible increase in the load-carrying capacity of the system as a whole. Although when negatively skewed roughness occurs, the integrated impact of bearing deformation and slip velocity can be significantly minimized, it is made abundantly evident by the data that are graphically shown that this is not always the case. Even if the semi-vertical angle and aspect ratios are selected appropriately, it is still necessary to keep the slip parameter as low as possible in order to achieve any type of enhancing the bearing's capabilities. This is the case even if the parameter is kept as low as possible.

Keywords: Slip velocity, truncated conical plates, deformation, hydro magnetic fluid, roughness

Abbreviation

T. R.	Transverse Roughness
L. R.	Longitudinal Roughness

[a]rmpatel2711@gmail.com

L. C. C.	Load Carrying Capacity
L. B. C.	Load Bearing Capacity
S. F.	Squeeze Film
M. F.	Magnetic Field
N. D. T.	Non-dimensional Term
S. D.	Standard Deviation
M. C.	Morgan – Cameron
C & T	Christensen and Tonder's

Introduction

Experimental and theoretical works on the hydro magnetic lubrication of spongy and also plane metal bearings have been performed by a number of researchers. The conduction of lubricants across non-conductive, non-spongy surfaces using hydromagnetic S. F. bearings when there is a transverse magnetic field that has been investigated by Kuzma et al. (1964), Shukla (1965), Elco and Huges (1962), Dodge et al. (1965) and Kuzma (1964). The efficiency of a hydro magnetic S. F. among non-spongy surfaces has been studied by Shukla and Prasad (1965) and they have investigated the influence of surface conductivities on the output of the bearing system. The hydromagnetic impact on the spongy S. F. of annular plates was investigated by Sinha and Gupta (1974). Patel and Hingu (1978) observed this overall impact between circular discs for S. F. Patel and Gupta (1979) used the estimation of M - C and simplified this analysis for hydro magnetic S. F. efficiency among parallel plates with different geometrical shapes. The L.C. C. and time height relationship for S. F. among spongy plates was examined by Prakash and Vij (1973). Prajapati (1995) considered the output of a hydro magnetic S. F. for two spongy conical plates. Patel and Deheri (2007) proposed observing M. F. based S. F. behavior amid spongy conical plates. Conclusion was made that the enhancement of bearing system efficiency was depending on magnetic fluid as well as semi vertical angle of the cone.

Tzeng and Saibel (1967) studied a 2-dimensional (2-D) inclined slider bearing in the direction transverse to the sliding direction with 1-dimensional (1-D) irregularity by understanding the random presence of irregularity and introducing stochastic approaches. Many researchers (Davis, 1963; Christensen and Tonder, 1969a,b, 1970; Tonder, 1972; Berthe and Godet, 1973) have studied the impact of C & T on surface irregularity. The method of Tzeng and Saibel (1967) was developed and updated by C & T (Christensen and Tonder, 1969a,b, 1970) and suggested a systematic general analysis for both T. R. and L. R. Various of studies (Ting, 1975; Guha, 1993; Gupta and Deheri, 1996; Andharia, Gupta and Deheri, 1999) on C & T created the foundation for the investigation of the effects of surface irregularity.

Vadher et al. (2008) have examined and analyzed the hydro magnetic S. F. behavior between spongy transversely rough triangular plates. In the event of negatively skewed irregularity, the detrimental effect of sponginess and irregularity has been demonstrated to be somewhat offset by the favorable effect of hydro magnetization. With this goal in mind, an attempt was undertaken to explore the behavior of a hydro magnetic S. F. between spongy conducting transversely rough truncated conical plates.

Patel et al. (2015) investigated S. F. behavior of different spongy structures on rough conical plates. Recently, hydro magnetic S. F. in rough truncated conical plates having K. C. model based spongy structure was discussed by Adeshara et al. (2020).

Analysis

In a truncated conical plate bearing system, the top plate travels in the opposite direction as the bottom plate. The bearings have a compact, longitudinally rough surface that is protected by a porous outer layer. Below Figure is the bearing geometry. For a comprehensive analysis, refer Prakash and Vij (1973).

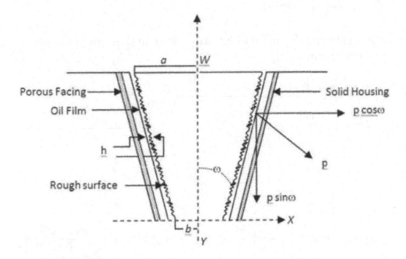

The film thickness is taken from C & T (Christensen and Tonder, 1969a,b, 1970) in accordance with Tzeng and Saibel (1967). Standard deviation, mean, and skewness details are available from Christensen and Tonder (Christensen and Tonder, 1969a,b, 1970). In this case, the equation of Reynolds' associate was stochastically averaged using the C & T approach (Christensen and Tonder, 1969a,b, 1970). Assuming axial symmetry of the magnetic fluid flow between the truncated conical plates, the amplitude of the slanted magnetic field *h* is a function of *x* that disappeared is *a*cosecω and *b*cosecω. For example,

$$p(a / \sin\omega) = 0 \; ; \; p(b / \sin\omega) = 0$$

where K has been selected to have the necessary strength of the magnetic field. Those proportions are suitable for the two side's measurements. As in Deheri et al. (2007), the angle of inclination of the M. F. with lower plate is determined. After performing a stochastic averaging on the equation in modified Reynolds form that governs film pressure p, we find that it is Prajapati's (1995) equation.

$$x^{-1}\frac{d}{dx}\left(x\frac{dp}{dx}\right) = \frac{h\mu A}{\left[\frac{2}{M^3}\left(\frac{M}{2}-\tanh\frac{M}{2}\right)-\frac{\psi}{\mu c^2}\right]} \cdot \frac{1}{\left[\frac{\varphi_0+\varphi_1+1}{\varphi_0+\varphi_1+\tanh\left(\frac{M}{2}\right)\times\left(\frac{2}{M}\right)}\right]}$$ (1)

Where

$$A = (h+p_a p'\delta)^{-3}\left[\frac{1+2sh}{1+4sh}\right]^{-3}\begin{vmatrix}1-\alpha(h+p_a p'\delta)^{-1}\left[\frac{1+2sh}{1+4sh}\right]^{-1/3}+6(h+p_a p'\delta)^{-2}\left[\frac{1+2sh}{1+4s}\right]^{-2/3}\\-10(h+p_a p'\delta)^{-3}(3\sigma^2\alpha+\alpha^3+\varepsilon)\left[\frac{1+2sh}{1+4sh}\right]^{-1}\end{vmatrix}$$

where p' is the standard pressure, and the corresponding conditions are,

$$p[a/\sin\omega] = 0\ ;\ p[b/\sin\omega] = 0$$ (2)

Solving the equation (1) using boundary condition, we can determine the pressure distribution using the form

$$p = \frac{-\dot{h}(a^2-b^2)\cosec^2\omega \cdot \left[\frac{\ln(x\sin\omega/b)}{\ln(a/b)}-\frac{(x\sin\omega/b)^2-1}{(a/b)^2-1}\right]B}{4\left[2\times M^{-3}\left(\frac{M}{2}-\tanh\frac{M}{2}\right)-\frac{\psi}{\mu c^2}\right]} \cdot \frac{1}{\left[\frac{\varphi_0+\varphi_1+1}{\varphi_0+\varphi_1+\tanh(M/2)\times(M/2)^{-1}}\right]}$$

Where

$$B = H^{-3}(1+\bar{p}\bar{\delta})\left[\frac{1+2\bar{s}}{1+4\bar{s}}\right]^{-3}$$

$$\cdot\begin{vmatrix}1-\bar{\alpha}(1+\bar{p}\bar{\delta})^{-1}\left[\frac{1+2\bar{s}}{1+4\bar{s}}\right]^{-1/3}+6(1+\bar{p}\bar{\delta})^{-2}(\bar{\sigma}^2+\bar{\alpha}^2)\left[\frac{1+2\bar{s}}{1+4\bar{s}}\right]^{-2/3}\\-10(1+\bar{p}\bar{\delta})^{-3}(3\bar{\sigma}^2\bar{\alpha}+\bar{\alpha}^3+\varepsilon)\left[\frac{1+2\bar{s}}{1+4\bar{s}}\right]^{-1}\end{vmatrix}$$

The distribution of N. D. T. pressure is obtained

$$P = \frac{-ph^3}{\mu\pi\dot{h}(a^2-b^2)\cosec\omega}$$

$$= \frac{\cos ec\omega\cdot\left[\frac{\ln\left(\frac{x\sin\omega}{b}\right)}{\ln\left(\frac{a}{b}\right)}-\frac{(x\sin\omega/b)^2-1}{(a/b)^2-1}\right]B}{4\pi\left[2\times M^{-3}\left(\frac{M}{2}-\tanh\frac{M}{2}\right)-\frac{\psi}{c^2}\right]} \cdot \frac{1}{\left[\frac{\varphi_0+\varphi_1+1}{\varphi_0+\varphi_1+\tanh(M/2)\times(M/2)^{-1}}\right]}$$

Where in

$$B = H^{-3}(1+\bar{p}\bar{\delta})^{-3}\left[\frac{1+2\bar{s}}{1+4\bar{s}}\right]^{-3}$$

$$\left| \begin{array}{c} 1 - \bar{\alpha}(1 + \bar{p}\bar{\delta})^{-1} \left[\frac{1+2\dot{s}}{1+4\bar{s}}\right]^{-1/3} + 6(1 + \bar{p}\bar{\delta})^{-2}(\bar{\sigma}^2 + \bar{\alpha}^2)\left[\frac{1+2\dot{s}}{1+4\bar{s}}\right]^{-2/3} \\ -10(1 + \bar{p}\bar{\delta})^{-3}(3\bar{\sigma}^2\bar{\alpha} + \bar{\alpha}^3 + \bar{\varepsilon})\left[\frac{1+2\dot{s}}{1+4\bar{s}}\right]^{-1} \end{array} \right|$$

Where

$$\psi = \phi H_0 h^{-3}, \bar{\sigma} = \sigma h^{-1}, \bar{\alpha} = \alpha x h^{-1}, \bar{\varepsilon} = \varepsilon x h^{-3}, \bar{p} = p_a p', \bar{\delta} = \frac{\delta}{h}, \bar{s} = sh$$

Then the load given by

$$w = 2\pi \int_{b\cos ec\omega}^{a\cos ec\omega} p \cdot x dx$$

is represented in dimensions as

$$w = \frac{-\dot{h}\pi(a^2 - b^2)\cos ec^4\omega \cdot \left[(a^2 + b^2) - \frac{(a^2-b^2)}{\ln\left(\frac{a}{b}\right)}\right] B}{8\left[2 \times \left(\frac{M}{2} - \tanh\frac{M}{2}\right)M^{-3} - \frac{\psi}{c^2}\right]} \cdot \frac{1}{\left[\frac{\phi_1+\phi_0+1}{\phi_1+\phi_0+\tanh(M/2)\times(2/M)}\right]}$$

The load using dimensionless terms is

$$W = -\frac{wh^3}{\mu \dot{h}\pi^2(a^2-b^2)^2\cos ec^2\omega} = \frac{\cos ec^2\omega \cdot \left[\frac{(a/b)^2+1}{(a/b)^2-1} \frac{1}{\ln(a/b)}\right] B}{8\pi\left[2\times M^{-3}\left(\frac{M}{2}-\tanh\frac{M}{2}\right)-\frac{\psi}{c^2}\right]} \cdot \frac{1}{\left[\frac{\phi_1+\phi_0+1}{\phi_1+\phi_0+\tanh(M/2)\times(2/M)}\right]}$$

Results and discussions

It is clear that equation (3) gives the N. D. T. pressure distribution, whereas the dimensionless load is calculated by using equation (4). The influence of conductivity factors on pressure and load distribution is estimated by

$$\frac{\phi_1+\phi_0+\left(\tanh\frac{M}{2}\right)/\left(\frac{M}{2}\right)}{\phi_1+\phi_0+1}$$

which turns to

$$\frac{\phi_1+\phi_0}{\phi_1+\phi_0+1}$$

Since, $\tanh(M) \cong 1$ and $\frac{2}{M} \cong 0$ for large values of M. Furthermore, the pressure and load carrying capacity rise as $\phi_0 + \phi_1$ increases, since both function are increasing functions of $\phi_1 + \phi_0$.

Figures (12.1–12.10) shows fluctuation of load considering different values of the magnetization parameter $\phi_1 + \phi_0$, s*, a*, e*, k, w, s*, y, d* and p*. These graphs clearly shows that the load rises significantly with regard to the magnetization parameter, with the influence of (-ve) a* being the most prominent, followed by angle w. Furthermore, the initial influence of the standard deviation is essentially small, and the increased load for the use of M and s* is substantially less than that of the other situation. Moreover, the effect of S.D. is rather negative. Load distribution versus conductivity $\phi_1 + \phi_0$ for different values of the parameters s*, a*, e*, k, w, s*, y, d*, and p* are depicted in Figures (12.11–12.19). It has been discovered the load tends to keep on growing with conductivity, with the rate of growth being substantially faster in early phases. In this case the combined effect

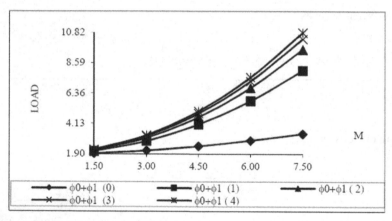

Figure 12.1 Load of M & $\phi_0 + \phi_1$

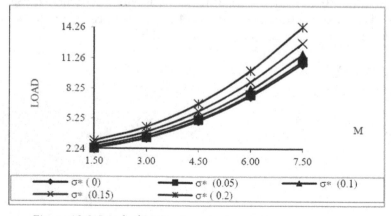

Figure 12.2 Load of M & σ*

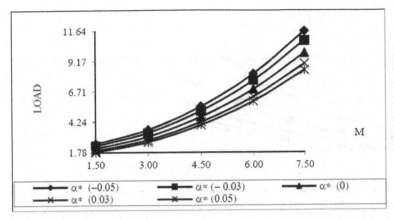

Figure 12.3 Load of M & α*

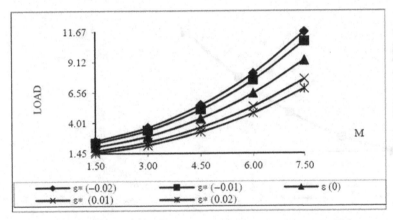

Figure 12.4 Load of M & ε*

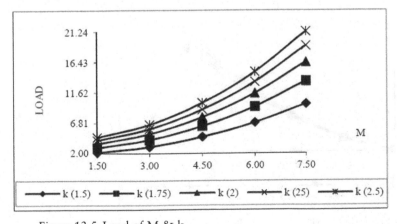

Figure 12.5 Load of M & k

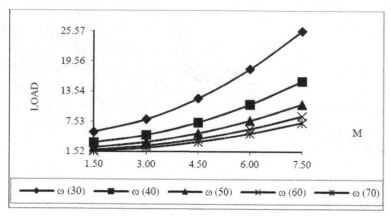

Figure 12.6 Loadof M & ω

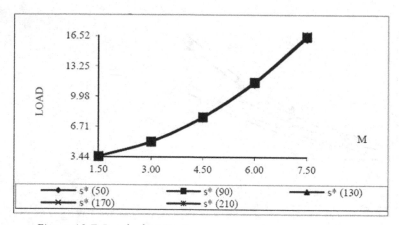

Figure 12.7 Load of M & s*

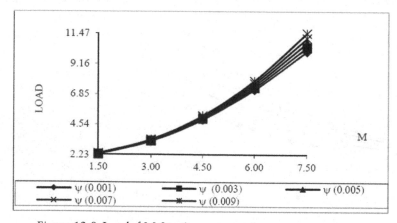

Figure 12.8 Loadof M & ψ*

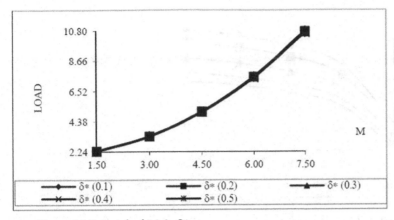

Figure 12.9 Load of M & δ*

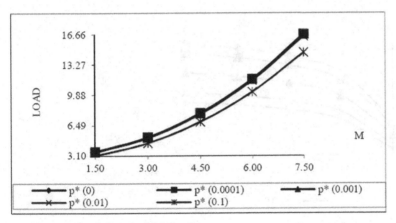

Figure 12.10 Load of M & p*

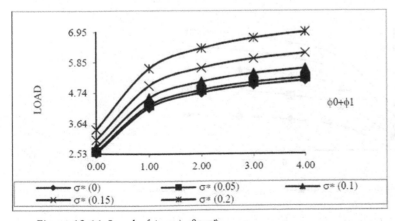

Figure 12.11 Load of $\phi_0 + \phi_1$ & σ*

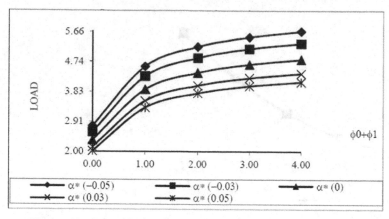

Figure 12.12 Load of $\phi_0 + \phi_1$ & α^*

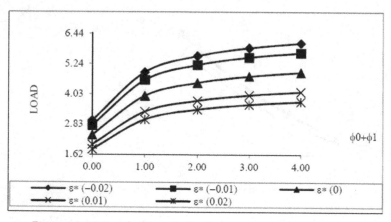

Figure 12.13 Load of $\phi_0 + \phi_1$ & ε^*

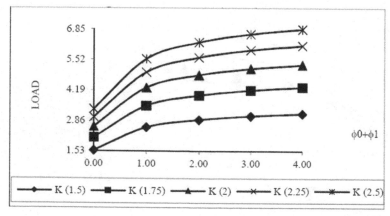

Figure 12.14 Load of $\phi_0 + \phi_1$ & k

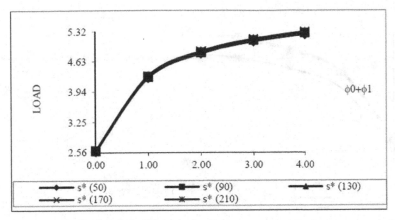

Figure 12.15 Load of $\phi_0 + \phi_1$ & s*

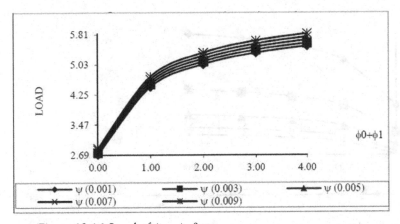

Figure 12.16 Load of $\phi_0 + \phi_1$ & ψ

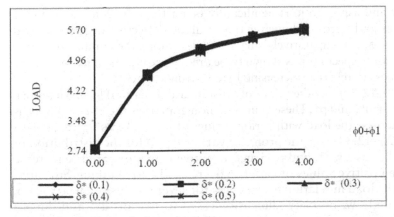

Figure 12.17 Load of $\phi_0 + \phi_1$ & δ*

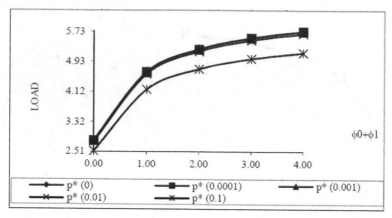

Figure 12.18 Load of $\phi_0 + \phi_1$ & p^*

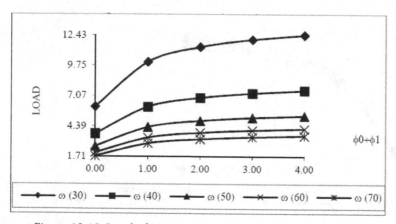

Figure 12.19 Load of $\phi_0 + \phi_1$ & ω

of conductivity and aspect ratio is significantly better than the combined impact of conductivities and porosity, however the combined effect of negative e^* and conductivity $\phi_1 + \phi_0$ is comparatively smaller. Furthermore, the combined impact of conductivity and aspect ratio is shown to be among the effects of negative variance and skewness in terms of increasing load bearing capacity.

Figures (12.20–12.27) give the profile of σ^* (standard deviation) for the load w. r. t. $a^*, e^*, k, w, s^*, y, d^*$, and p^*. These figures demonstrate that the effect of S. D. is considerably increase the load with various values of $a^*, e^*, k, ws^*, y^*, d^*$, and p^*.

Figures (12.28–12.34) give the profile of variance (a^*) for the distribution of load w. r. t. e^*, k, w, s^*, y, d^*, and p^*. Hence, negative values of variance increase the load, while positive values of variance decrease the load taking capacity.

Changes in the load in relation to skewness for several values of aspect ratio, semi vertical angle w, porosity, dimensionless deformation d^*, and dimensionless pressure p^* is presented in Figures (12.34–12.40), respectively. Here

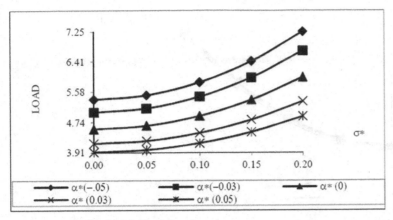

Figure 12.20 Load of σ* & α*

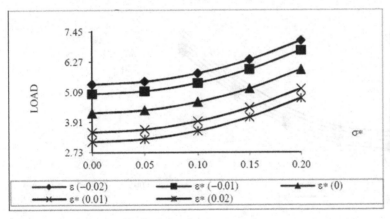

Figure 12.21 Load of σ* & ε*

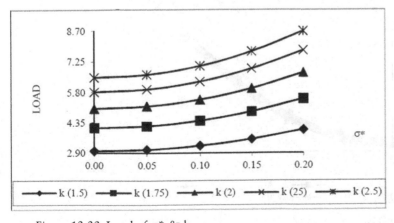

Figure 12.22 Load of σ* & k

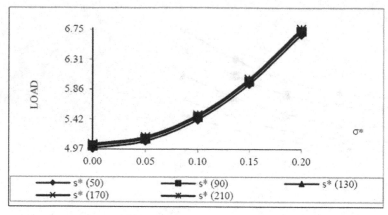

Figure 12.23 Load of σ* & s*

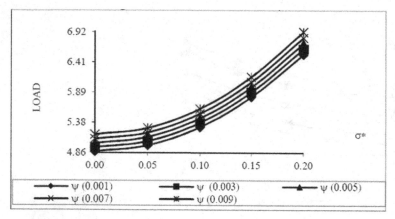

Figure 12.24 Load of σ* & ψ

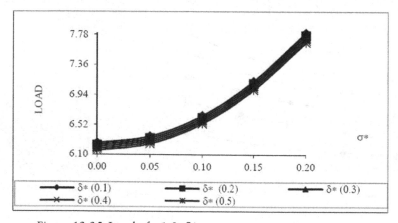

Figure 12.25 Load of σ* & δ*

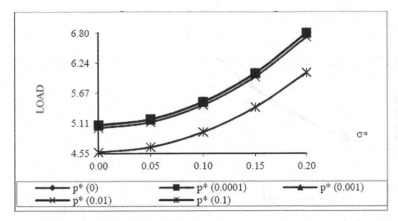

Figure 12.26 Load of σ* & p*

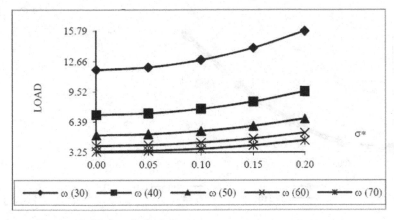

Figure 12.27 Load of σ* & ω

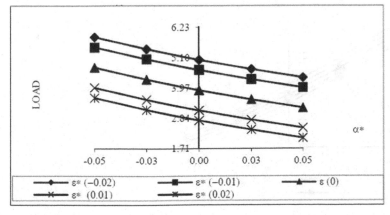

Figure 12.28 Load of α* & ε*

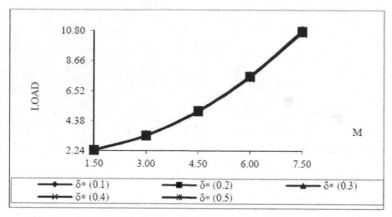

Figure 12.29 Load of α* & k

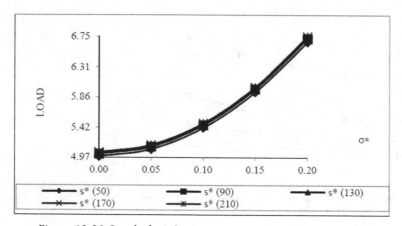

Figure 12.30 Load of α* & s*

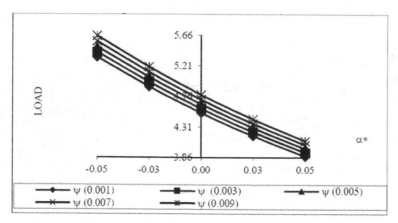

Figure 12.31 Load of α* & ψ

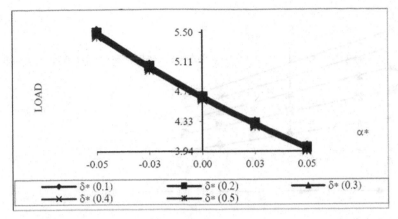

Figure 12.32 Load of α* & δ*

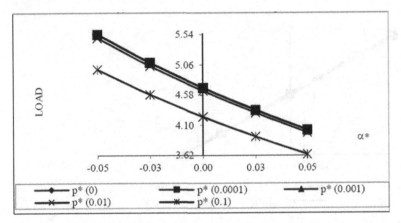

Figure 12.33 Load of α* & p*

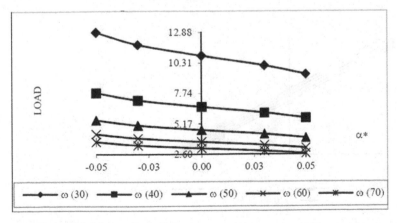

Figure 12.34 Load of α* & ω

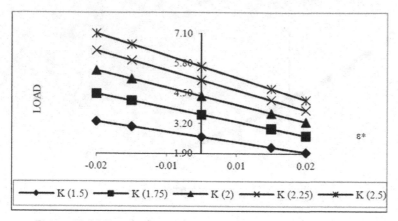

Figure 12.35 Load of ε* & k

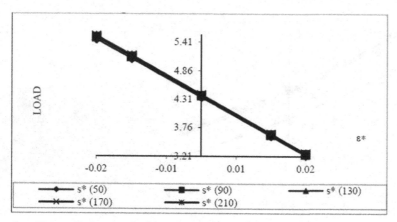

Figure 12.36 Load of ε* & s*

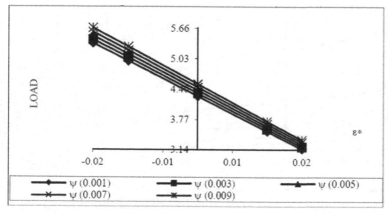

Figure 12.37 Load of ε* & ψ

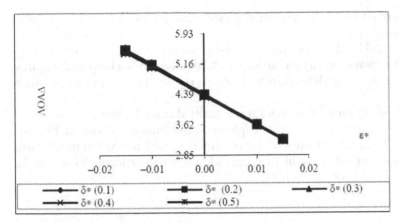

Figure 12.38 Load of ε^* & δ^*

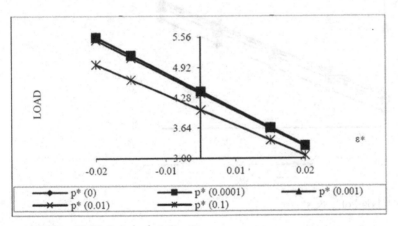

Figure 12.39 Load of ε^* & p^*

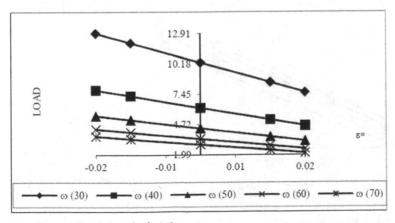

Figure 12.40 Load of ε^* & ω

smaller values of (-) variance increase the load while, (+) variance decrease the load taking capacity.

In Figures (12.41–12.45) it can be easily shown that the load sharply increases w. r. t. various values of k, w, s*, d*, and p*. However, Figures (12.46–12.49) shows the deformation decrease the load with respect to other parameters.

The load of aspect ratio increases substantially due to the slip parameter s*, ambient pressure parameter p* and angle w of the truncated cone in Figures (12.50–12.52). The higher values of slip parameter and angle w fail to increase the load when the pressure ambient is considered for the specified value as in Figures 12.53 and 12.54.

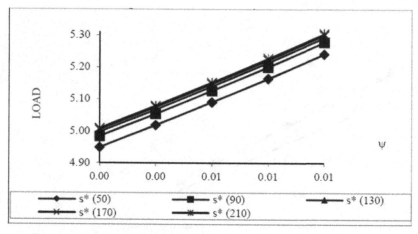

Figure 12.41 Load of ψ & s*

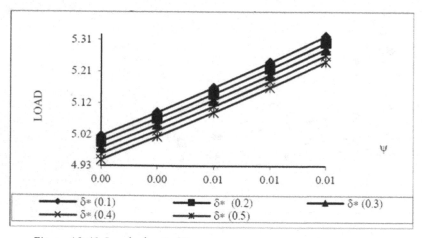

Figure 12.42 Load of ψ & δ*

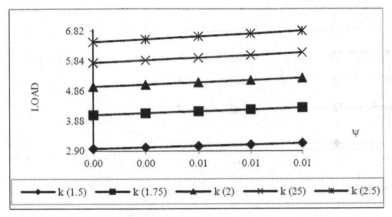

Figure 12.43 Load of ψ & k

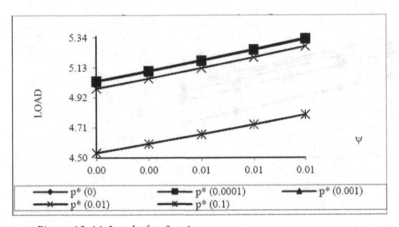

Figure 12.44 Load of ψ & p*

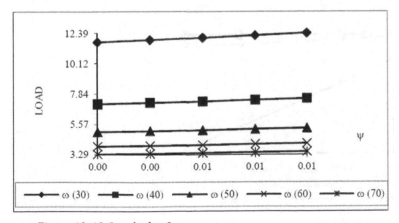

Figure 12.45 Load of ψ & ω

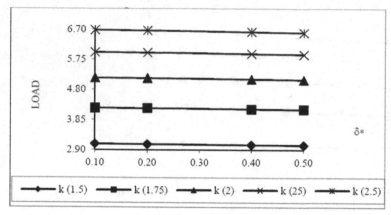

Figure 12.46 Load of δ & k

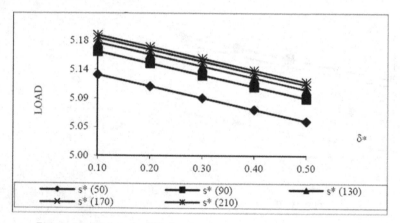

Figure 12.47 Load of δ* & s*

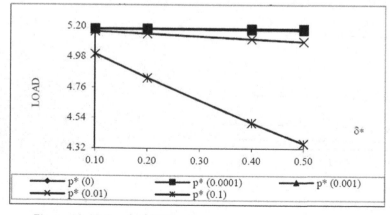

Figure 12.48 Load of δ* & p*

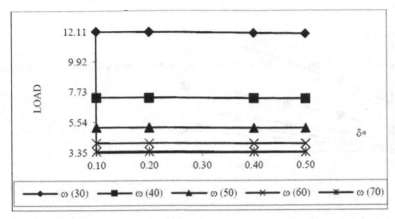

Figure 12.49 Load of δ* & ω

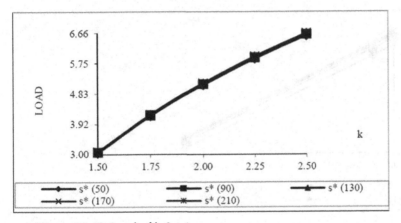

Figure 12.50 Load of k & s*

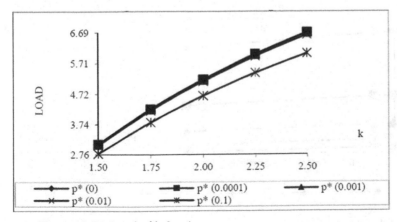

Figure 12.51 Load of k & p*

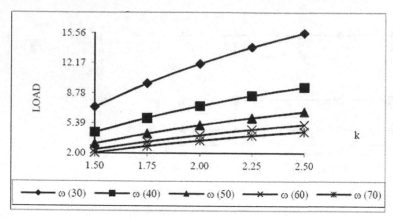

Figure 12.52 Load of k & ω

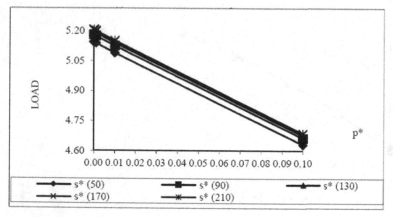

Figure 12.53 Load of p* & s*

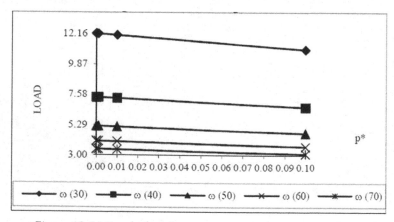

Figure 12.54 Load of p* & ω

Conclusion

- This study suggests that in the situation of negatively skewed irregularity, hydromagnetic lubrication can significantly reduce the negative effects of deformation and slip velocity because the standard deviation associated with irregularity provides higher load.
- Even when considerable deformation and slip velocity are present, compared to longitudinal irregularity patterns, the situation is still substantially better.
- Even if the slip value is at a minimal, the deformation and irregularity features must be taken into account while building the bearing system.

References

1. Adeshara, J. V., Patel, H. P., Deheri, G. M., and Patel, R. M. (2020). Theoretical study of hydromagnetic squeeze film rough truncated conical plates with Kozeny-Carman model based spongy structure. *Proc. Engg. Sci.*, 2(4), 389–400.
2. Andharia, P. I., Gupta, J. L., and Deheri, G. M. (1999). Effect of transverse surface roughness on the behaviour of squeeze film in a spherical bearing. *J. Appl. Mec. Engg.*, 4, 19–24.
3. Berthe, D. and Godet, M. (1973). A more general form of Reynolds equation – Application to rough surfaces. *Wear*, 27, 345–357.
4. Christensen, H. and Tonder, K. C. (1969a).Tribology of rough surfaces: Parametric study and comparison of lubrication models. SINTEF Report No. 22/69-18.
5. Christensen, H. and Tonder, K. C. (1969b). Tribology of rough surfaces: Stochastic models of hydrodynamic lubrication. SINTEF Report No. 10/69-1.
6. Christensen, H. and Tonder, K. C. (1970). Tribology of rough surfaces: A stochastic model of mixed lubrication. SINTEF Report No. 18/70-21.
7. Davis, M. G. (1963). The generation of pressure between rough lubricated, moving deformable surfaces. *Lub. Engg.*, 19, 246–252.
8. Dodge, F. T., Osterle, J. F., and Rouleau, W. T. (1965). Magnetohydrodynamic squeeze film bearings. *J. Basic Engg. Trans. ASME*, 87, 805–809.
9. Elco, R. A. and Huges, W. F. (1962). Magnetohydrodynamic pressurization in liquid metal lubrication. *Wear*, 5, 198–212.
10. Guha, S. K. (1993). Analysis of dynamic characteristics of hydrodynamic journal bearings with isotropic roughness effects. *Wear*, 167, 173–179.
11. Gupta, J. L. and Deheri, G. M. (1996). Effect of roughness on the behaviour of squeeze film in a spherical bearing. *Tribol. Trans.*, 39, 99–102.
12. Irmay, S. (1995). Flow of liquid through cracked media. *Bull. Res. Council Isr.*, 5A(1), 84.
13. Kuzma, D. C. (1964). Magnetohydrodynamic squeeze films. *J. Basic Engg. Trans. ASME*, 86, 441–444.
14. Kuzma, D. C., Maki, E. R., and Donnelly, R. J. (1964). The magnetohydrodynamic squeeze films. *J. Fluid Mec.*, 19, 395–400.
15. Majumdar. B. C. (1985). *Introduction to Tribology of Bearings*. Wheeler Co. Ltd., India: Wheeler Publisher.
16. Patel, K. C. and Gupta, J. L. (1979). Behavior of hydromagnetic squeeze film between porous plates. *Wear*, 56, 327–339.
17. Patel, K. C. and Hingu, J. V. (1978). Hydromagnetic squeeze film behavior in porous circular disks. *Wear*, 49, 239–246.

18. Patel, R. M. and Deheri, G. M. (2007). Magnetic fluid based squeeze film between porous conical plates. *Indus. Lub. Tribol.*, 59(3), 143–147.
19. Patel, R. M., Deheri, G. M., and Vahder, P. A. (2015). Hydromagnetic rough spongy circular step bearing. *Eastern Acad. J.*, 3, 71–87.
20. Prajapati, B. L. (1995). On Certain Theoretical Studies in Hydrodynamic and Electromagnetohydrodynamic Lubrication. Ph.D. Thesis, Sardar Patel University, VallabhVidyanagar, Gujarat, India.
21. Prakash, J. and Vij, S. K. (1973). Load capacity and time height relations for squeeze film between porous plates. *Wear*, 24, 309–322.
22. Shukla, J. B. (1965). Hydromagnetic theory of squeeze films. *ASME*, 87, 142–144.
23. Shukla, J. B. and Prasad, R. (1965). Hydromagnetic squeeze films between two conducting surfaces. *J. Basic Engg Trans. ASME*, 87, 818–823.
24. Sinha, P. C. and Gupta, J. L. (1974). Hydromagnetic squeeze films between porous annular disks. *J. Math. Phy. Sci.*, 8, 413–422.
25. Ting, L. L. (1975). Engagement behaviour of lubricated porous annular disks Part I: Squeeze film phase, surface roughness and elastic deformation effects. *Wear*, 34, 159–172.
26. Tonder, K. C. (1972). Surface distributed waviness and roughness. *First World Conf. Indus. Tribol.*, A3, 128.
27. Tzeng, S. T. and Saibel, E. (1967). Surface roughness effect on slider bearing lubrication. *J. Lub. Technol. Trans. ASME*, 10, 334–348.
28. Vadher, P. A., Deheri, G. M., and Patel, R. M. (2008). Hydromagnetic squeeze film between conducting porous transversely rough triangular plates. *Ann. Faculty Engg. Hunedoara*, 6, 155–168.

13 Influence of the hydro-magnetic squeeze film on a longitudinally rough circular step bearing: Comparison of porous structures

Dr. Jatinkumar V. Adeshara[1,a], Dr. Suresh G. Sorathiya[2,b], Dr. Hardik P. Patel[3], Dr. Gunamani B. Deheri[4] and Rakesh M. Patel[5]

[1]Assistant Professor, Vishwakarma Government Engineering College, Chandkheda, Ahmedabad-382424, Gujarat, India

[2]Assistant Professor, Department of Science, Sal Institute of Technology & Engineering Research, Ahmedabad, Gujarat, India

[3]Assistant Professor, Department of Humanity and Science, L. J. Institute of Engineering and Technology, LJK University, Ahmedabad, Gujarat, India

[4]Retired Associate Professor, Department of Mathematics, S. P. University, Vallabh Vidyanagar-388120, Gujarat, India

[5]Assistant Professor, Department of Mathematics, Gujarat Arts and Science College, Ahmedabad-380006, Gujarat, India

Abstract

This study compares the effects of spongy structures based on the Kozeny–Carmon (K. C.) model (a globular sphere model) and Irmay's (I. M.) model (a capillary fisher's model) structures, which are modeled on hydro-magnetic based squeeze film in irregular circular step bearing. C & T stochastically averaging model have been adopted to determine the role of longitudinal irregularity. The concerned generalized stochastically averaged Reynolds' equation is solved with feasible boundary conditions to get the distribution of pressure. Further, the load carrying capacity (L. C. C.) is computed in consideration of suitable conditions. The outcomes depict that increasing magnetization parameter leads to increased load. Also, the impact of longitudinal roughness has been discussed for both the structures. However, this effect is sharper in the case of Irmay's model. The adverse effect of irregularity and sponginess can be weakened by the (+ve) effect of magnetization at least in the case of G. S. model.

Keywords: Hydro-magnetic squeeze film, longitudinal roughness, circular step bearing, porous structures

Abbreviation

K. C. Kozeny–Carmon Model (a globular sphere model)
C & T Christensen and Tonder
G. S. Globular Sphere Model
I. M. Irmay's Model (a capillary fisher's model)

[a]desharajatin01@gmail.com, [b]suresh.sorathiya@sal.edu.in

L. C. C. Load Carrying Capacity
L. T. C. Load Taking Capacity
T. R.   Transverse Roughness
L. R.   Longitudinal Roughness
S. F.   Squeeze Film
M. F.   Magnetic Field
B. C.   Boundary Conditions
M. C.   Morgan–Cameron
S. D.   Standard Deviation
N. D.   Non-Dimensional

Introduction

Experimental and theoretical studies on the hydro-magnetic lubrication of spongy and also metal bearings which is plane have been performed by a number of researchers (Dodge et al., 1965; Kuzma et al., 1964; Elco and Huges, 1962; Kuzma, 1964). Shukla (1965) worked with hydro-magnetic S. F. bearings in the transverse magnetic field's presence for the conduction of lubricants among non-conductive non-spongy surfaces. Shukla and Prasad (1965) addressed the efficiency of a hydro-magnetic S. F. among non-spongy surfaces and investigated the influence of surface conductivities on the output of the bearing. The hydro-magnetic impact on the spongy S. F. of annular plates was investigated by Sinha and Gupta (1974). Patel and Hingu (1978) observed this overall circular disc's impact for S. F. Patel and Gupta (1979) used the approximation of M. C. and simplified this investigation, for hydro-magnetic S. F. efficiency among parallel plates with different geometrical shapes. The load and time height relationship for S. F. among spongy plates was examined by Prakash and Vij (1973). The output of a hydro-magnetic S. F. for two spongy conical plates was regarded by Prajapati (1995). The observation of M. F. based S. F. behavior among spongy conical plates was proposed by Patel and Deheri (2007). Here, it was inferred that the cone's M. F. and semi-vertical angle performed key role in improving the bearing efficiency.

Tzeng and Saibel (1967) studied a 2-dimensional (2-D) inclined slider bearing with 1-dimensional (1-D) irregularity by understanding the arbitrary presence of irregularity and introducing stochastic approaches. Many researcher's C & T (Davis, 1963; Christensen and Tonder, 1969a,b, 1970; Tonder, 1972; Berthe and Godet, 1973) have calculated the impact of surface irregularity. The method of Tzeng and Saibel (1967) was established and updated by C & T (Christensen and Tonder, 1969a,b, 1970) and proposed a systematic general scrutiny for both T. R. and L. R. In a several of studies (Ting, 1975; Guha, 1993; Gupta and Deheri, 1996; Andharia, Gupta and Deheri, 1999), C & T created the foundation for the investigation of the effects of surface irregularity.

The hydro-magnetic S. F. behavior among the spongy transversely rough triangular plates has recently studied and analyzed by Vadher et al. (2008). The (-) impact of sponginess and irregularity has been shown to be defused to some degree by the (+) impact of hydro magnetization in the case of (-) skewed irregularity. An attempt was made here to investigate the action of a hydro-magnetic

S. F. among spongy conducting transversely rough truncated conical plates with this end in view.

Several spongy structures' S. F. behavior on rough conical plates was studied by Patel et al. (2015). Adeshara et al. (2020) have just discussed hydro-magnetic S. F. in rough truncated conical plates with a spongy structure based on the K. C. model.

Analysis

Shape of bearing is given in below Figure - I

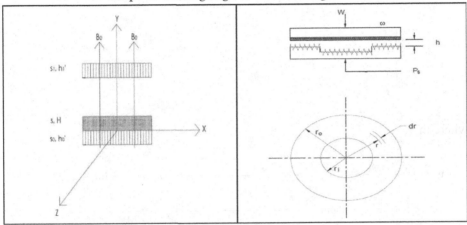

The surfaces are rough (longitudinally). The film thickness is brought from Christensen and Tonder (Christensen and Tonder, 1969a,b, 1970). The pressure for a circular step bearing is calculated using Reynolds' type as (Vadher, 2008; Majumdar, 1985; Patel and Deheri, 2007)

$$Q = -\frac{2\pi r \dfrac{dp}{dr} \left[\dfrac{2}{M^3 A}\left[\dfrac{M}{2} - \tanh \dfrac{M}{2} \right] + \dfrac{\psi l_1}{c^2 A} \right] \left[\dfrac{\phi_0 + \phi_1 + 1}{\phi_0 + \phi_1 + \left(\tanh \dfrac{M}{2} \right) \Big/ \left(\dfrac{M}{2} \right)} \right]}{12 \mu} \tag{1}$$

where

$$A = h^{-3}[\, 1 - 3\alpha h^{-1} + 3h^{-2}(\sigma^2 + \alpha^2) - \frac{3}{40}(\varepsilon + 3\sigma^2 \alpha + \alpha^3)\,]$$

ψ is the spongy structure of the spongy region, l_1 is the width of spongy facing.

Case – I (Figure I displays G. S. model)
The G. S. particles are supposed to fill the spongy substance in this model. Dc is the average particle size. According to the findings of K. C. the permeability of the spongy layer is

$$\psi = \frac{D_c^2 e_1^{\,3}}{180(1-e_1)^2}$$

where e_1 is the sponginess.

Integrate the given equation (1) in terms of B. C.

$$P(r_0) = 0 \text{ and } p(r_i) = p_s \tag{2}$$

The following is the leading equation of the pressure p of film region:

$$p = p_s \frac{\ln\left(\dfrac{r}{r_o}\right)}{\ln\left(\dfrac{r_i}{r_o}\right)}$$

where in

$$P_s = \frac{6}{\pi\left[\dfrac{2}{M^3 A}\left[\dfrac{M}{2} - \tanh\dfrac{M}{2}\right] + \dfrac{\psi\, l_1}{c^2 A}\left[\dfrac{\phi_0 + \phi_1 + 1}{\phi_0 + \phi_1 + \left(\tanh\dfrac{M}{2}\right)\Big/\left(\dfrac{M}{2}\right)}\right]\right]} \ln\left(\dfrac{r_o}{r_i}\right)$$

Introducing the non-dimensional terms

$$P_s^* = \frac{6\ln\left(\dfrac{1}{k}\right)}{\pi\left[\dfrac{2}{M^3 B}\left[\dfrac{M}{2} - \tanh\dfrac{M}{2}\right] + \dfrac{\overline{\psi}\, e_1^3}{15B(1-e_1)^2 c^2}\left[\dfrac{\phi_0 + \phi_1 + 1}{\phi_0 + \phi_1 + \left(\tanh\dfrac{M}{2}\right)\Big/\left(\dfrac{M}{2}\right)}\right]\right]} \tag{3}$$

$$B = m\;(h)\;\; h^3 = \left(1 + 3\times(\sigma^{*2} + a^{*2}) - \frac{3}{40}\times(3\times\sigma^{*2}\times a^* + \varepsilon^* + a^{*3}) - 3\times a^*\right)$$

$$a^* = \left(\frac{h}{a}\right)^{-1} \qquad \sigma^* = \left(\frac{h}{\sigma}\right)^{-1} \qquad \varepsilon^* = \left(\frac{h^3}{\varepsilon}\right)^{-1} \qquad k = \left(\frac{r_i}{r_o}\right)$$

$$\overline{\psi} = \frac{D_c^2 l_1}{h_0^3} \qquad \overline{h} = \left[\frac{h_0}{h}\right]^{-1} \qquad \overline{h}_1 = \frac{h_1}{h_0} \qquad \overline{h}_2 = \frac{h_2}{h_0}$$

$$\sigma^* = \frac{\sigma}{h_0} \qquad a^* = \frac{a}{h_0} \qquad \varepsilon^* = \frac{\varepsilon}{h_0^3}$$

We can get load w by integrating pressure in dimensionless form as

$$W = \frac{P_s{}'\left(2\ln\left(\frac{1}{k}\right)\right)^{-1}}{(1-k^2)^{-1}}$$ (4)

Case – II (Figure II shows C. F. model)

Three sets of cracks that are mutually orthogonal are included in this model (average solid size Ds). Irmay (1955) calculated the equation for the spongy structural parameter assuming no loss of hydraulic gradient at the connection.

$$\psi = \frac{\left(\dfrac{1-m}{}\right)^{2/3} D_s^{\,2}}{12 * m}$$

where m = 1 - e_1, e_1 being the porosity.

The following is the leading equation of the pressure p of film region:

$$p = p_s \frac{\ln\left(\dfrac{r}{r_o}\right)}{\ln\left(\dfrac{r_i}{r_o}\right)}$$

$$P_s = \frac{6}{\pi\left[\dfrac{2}{M^3 A}\left[\dfrac{M}{2} - \tanh\dfrac{M}{2}\right] + \dfrac{\psi l_1}{c^2 A}\right]\left[\dfrac{\phi_o + \phi_1 + 1}{\phi_o + \phi_1 + \left(\tanh\dfrac{M}{2}\right)\left(\dfrac{M}{2}\right)}\right]} = \ln\left(\dfrac{r_o}{r_i}\right)$$

Using demesnes quantities

$$P_s{}^* = \frac{6\ln\left(\dfrac{1}{k}\right)}{\pi\left[\dfrac{2}{M^3 B}\left[\dfrac{M}{2} - \tanh\dfrac{M}{2}\right] + \dfrac{\psi\,(1-m)^{2/3}}{Bmc^2}\right]\left[\dfrac{\phi_o + \phi_1 + 1}{\phi_o + \phi_1 + \left(\tanh\dfrac{M}{2}\right)\left(\dfrac{M}{2}\right)}\right]}$$ (5)

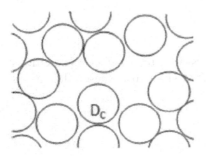

Figure – II K. C.'s spongy structure model

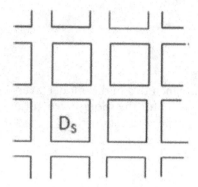

Figure – III Irmay's spongy structure model

—◆— e=0.2   —■— e=0.3   —▲— e=0.4   —✕— e=0.5   —✳— e=0.6

—◆— ψ=0   —■— ψ=0.005   —▲— ψ=0.01   —✕— ψ=0.015   —✳— ψ=0.020

——◆— k=0.35   —■— k=0.45   —▲— k=0.55   —✕— k=0.65   —✳— k=0.75

Graph – I Values in Graph

$$B = m\,(h) = h^{-3}\left(1 - 3\times\alpha*+3\times(\alpha*^2+\sigma*^2) - \frac{3}{40}\times(3\times\sigma*^2\times\alpha*+\varepsilon*+\alpha*^3)\right)$$

where

$$\overline{\psi} = \frac{D_s^2 l_1}{h_0^3}$$

The load w is determined by integrating the pressure, which has a non-dimensional term

$$W = \frac{P_s^{\cdot}(1-k^2)}{2\ln(k^{-1})} \tag{6}$$

Results and discussions

Equations (3) and (5) give the dimensionless pressure profile, whereas equations (4) and (6) guide the N. D. load distribution. It is obvious from these calculations that the load

$$W \propto P_s*$$

and

$$P_s = \cfrac{Q}{\pi\left[\dfrac{2}{M^3A}\left[\dfrac{M}{2}-\tanh\dfrac{M}{2}\right]+\dfrac{\psi\, l_1}{c^2A}\right]\left[\dfrac{\phi_0+\phi_1+1}{\phi_0+\phi_1+\left(\tanh\dfrac{M}{2}\right)\!/\!\left(\dfrac{M}{2}\right)}\right]}$$

- This demonstrates that given a fixed flow rate, as the stochastically averaged S. F. thickness decreases, the load rises.
- Consequently, when the flow velocity is maintained, the bearing demonstrates self-compensation.
- Equations (3–6) show how conductivity properties affect pressure and load distribution.

$$\frac{\phi_0+\phi_1+\left(\tanh\dfrac{M}{2}\right)\!/\!\left(\dfrac{M}{2}\right)}{\phi_0+\phi_1+1}$$

which turns to

$$\frac{\phi_0+\phi_1}{\phi_0+\phi_1+1}$$

since $\tanh(M)\cong 1$ and $(2/M)\cong 0$ for larger values of M. Furthermore, the pressure and load rise as $\phi 0+\phi 1$ grows since two of the functions are increasing functions of $\phi 0+\phi 1$.

Figures (13.1–13.12) depict the load distribution in relation to many parameters for K. C.'s G. S. model. Figures (13.13–13.24) depict the fluctuation of load in relation to Irmay's C. F. model. The load would therefore increase if the magnetization parameter were increased. Figures (13.1, 13.2, 13.13, 13.14) show that the

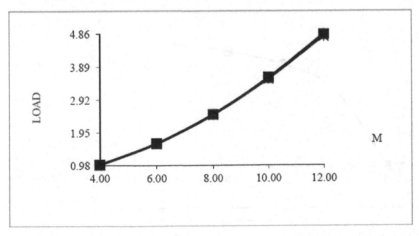

Figure 13.1 Load of M and e.

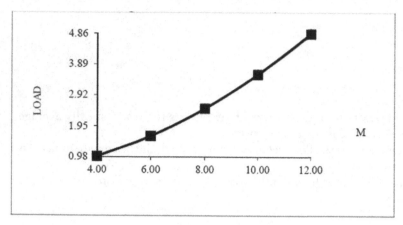

Figure 13.2 Load of M and ψ

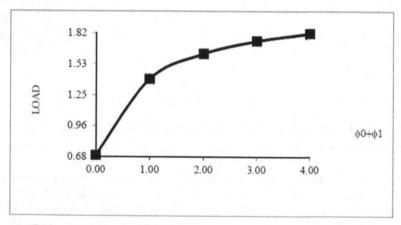

Figure 13.3 Load of φ0+φ1 and e

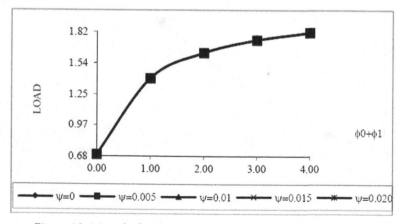

Figure 13.4 Load of φ0+φ1 and ψ

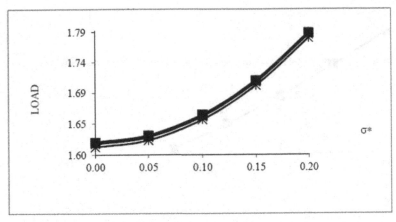

Figure 13.5 Load of σ* and e

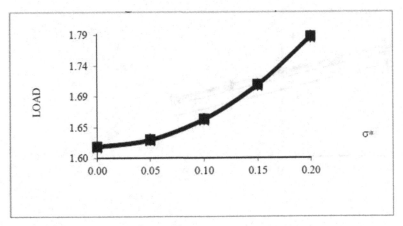

Figure 13.6 Load of σ* and ψ

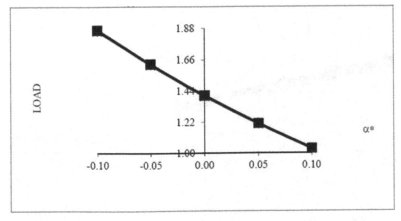

Figure 13.7 Load of α* and e

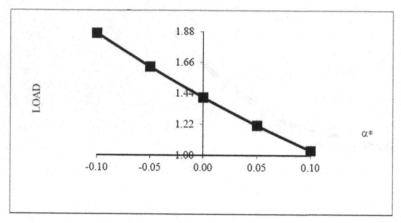

Figure 13.8 Load of α* and ψ

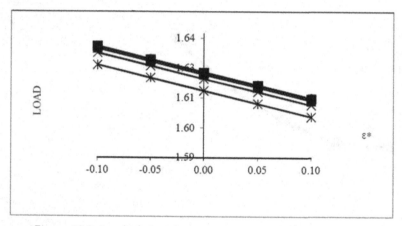

Figure 13.9 Load of ε* and e

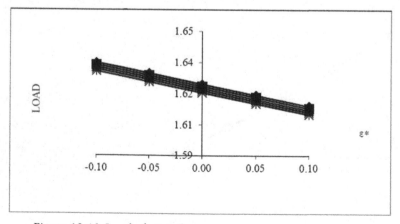

Figure 13.10 Load of ε* and ψ

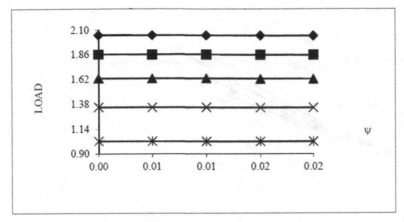

Figure 13.11 Load of ψ and k

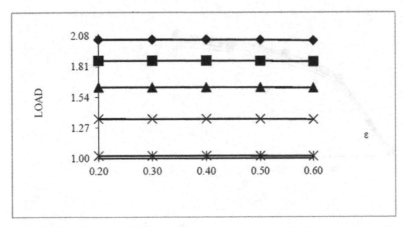

Figure 13.12 Load of e and k

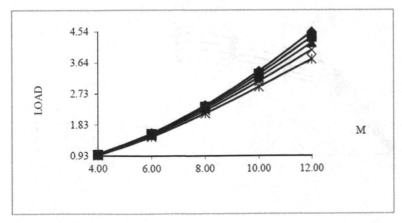

Figure 13.13 Load of M and e

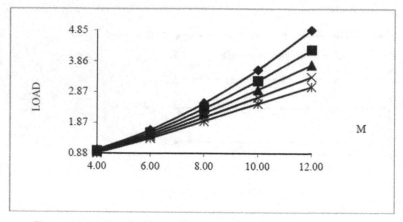

Figure 13.14 Load of M and ψ

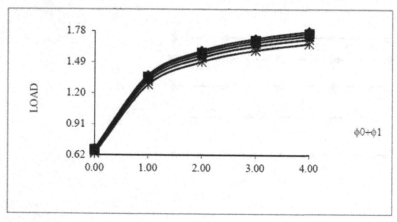

Figure 13.15 Load of φ0+φ1 and e

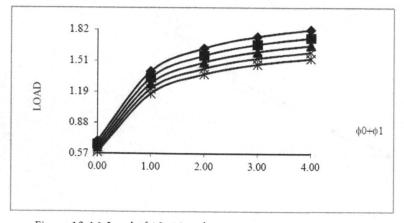

Figure 13.16 Load of φ0+φ1 and ψ

rate of change in load is considerably faster in the G. S. model of K. C. However, the spongy structure parameter's effect and sponginess on the discrepancy of load with regards to magnetization is insignificant up to certain extent in the case of G. S. model. Load increases sharply in case of both models.

Figures 13.3 and 13.4 shows the impact of $\phi 0 + \phi 1$ on the dist. of load with regards to K. C. model while the trend of load for Irmay's model is given in Figures 13.15 and 13.16. The rate of increase in the load is relatively more in G. S. model of K. C. compare to C. F. model of Irmay.

S. D. impact on the dist. of load with regards to both models are given in Figures 13.5, 13.6, 13.17, and 13.18. It is obvious that rising S. D. values lead to increased load, which in turn has a negative impact on S. F. performance. The impact of a spongy structure and sponginess on the variation of load w. r. t. S. D. is minimal in the case of the K. C. model, but it is minimal in the case of the Irmay's model, as can be seen from Figures 13.5 and 13.6.

It appears that the load for both models is reduced by the positive variance while the load grows as a result of variation (-ve). It's thrilling to see that for both the K. C. model (Figures 13.7 and 13.8) and the Irmay model (Figures 13.19 and 13.20), the effect of the spongy structure parameter on the fluctuation of load w.r.t. variance stays minimal. Figures 13.9, 13.10, 13.21 and 13.22 shows the effects of skewness for each model, respectively. The higher burden caused by variation (-ve) is exacerbated by negatively skewed irregularity. In this instance, the influence of the spongy structure parameter and sponginess is negligible for the K. C. model, although Irmay's model performs better.

Figures 13.11 and 13.12 shows that the combined outcome of the spongy structural parameter and the redii ratio (k) appears to be negative (23). The load reduction is, however, greatest at the beginning for Irmay's model. For the K. C. model and Irmay's model, the effects of sponginess and redii ratio are shown in Figures 13.12 and 13.24, respectively. When compared to the K. C. model, it is strictly lower in Irmay's model.

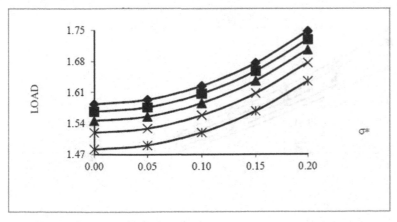

Figure 13.17 Load of σ^* and e

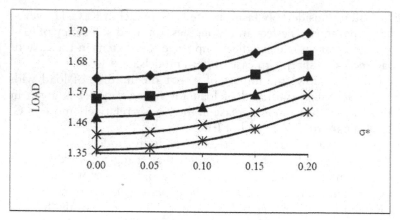

Figure 13.18 Load of σ* and ψ

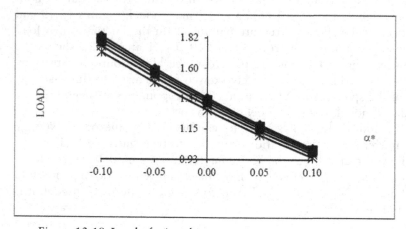

Figure 13.19 Load of α* and e

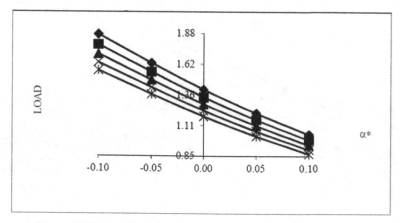

Figure 13.20 Load of α* and ψ

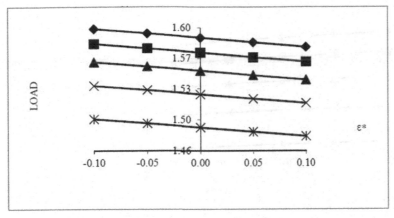

Figure 13.21 Load of ε^* and e

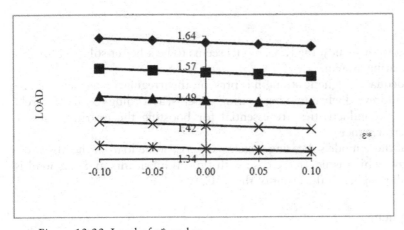

Figure 13.22 Load of ε^* and ψ

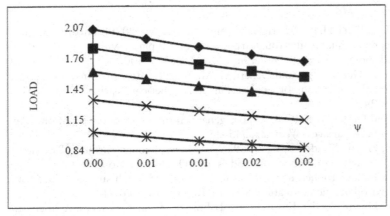

Figure 13.23 Load of ψ and k

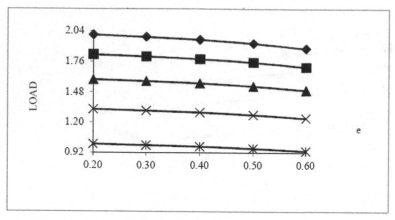

Figure 13.24 Load of e and k

Conclusion

- According to this study, the K. C. model seems to be a better solution for this kind of bearing system.
- Even if adequate magnetic strength is present, the irregularity aspect must be carefully addressed when making bearing systems, according to this research.
- Both plates' conductivities are essential for boosting the bearing system's overall performance.
- In both of these models' bearing systems support some load in the absenteeism of flow, which is never the case with typical lubricants and this load is considerably higher in the event of the K. C. type.

Acknowledgment

The reviewers' and editor's remarks are gratefully acknowledged by the authors.

References

1. Adeshara, J. V., Patel, H. P., Deheri, G. M., and Patel, R. M. (2020). Theoretical study of hydromagnetic squeeze film rough truncated conical plates with Kozeny-Carman model based spongy structure. *Proc. Engg. Sci.*, 2(4), 389–400.
2. Andharia, P. I., Gupta, J. L., and Deheri, G. M. (1999). Effect of transverse surface roughness on the behaviour of squeeze film in a spherical bearing. *J. Appl. Mec. Engg.*, 4, 19–24.
3. Berthe, D. and Godet, M. (1973). A more general form of Reynolds equation – Application to rough surfaces. *Wear*, 27, 345–357.
4. Christensen, H. and Tonder, K. C. (1969a).Tribology of rough surfaces: Parametric study and comparison of lubrication models. SINTEF Report No. 22/69-18.
5. Christensen, H. and Tonder, K. C. (1969b). Tribology of rough surfaces: Stochastic models of hydrodynamic lubrication. SINTEF Report No. 10/69-1.
6. Christensen, H. and Tonder, K. C. (1970). Tribology of rough surfaces: A stochastic model of mixed lubrication. SINTEF Report No. 18/70-21.

7. Davis, M. G. (1963). The generation of pressure between rough lubricated, moving deformable surfaces. *Lub. Engg.*, 19, 246–252.

8. Dodge, F. T., Osterle, J. F., and Rouleau, W. T. (1965). Magnetohydrodynamic squeeze film bearings. *J. Basic Engg. Trans. ASME*, 87, 805–809.

9. Elco, R. A. and Huges, W. F. (1962). Magnetohydrodynamic pressurization in liquid metal lubrication. *Wear*, 5, 198–212.

10. Guha, S. K. (1993). Analysis of dynamic characteristics of hydrodynamic journal bearings with isotropic roughness effects. *Wear*, 167, 173–179.

11. Gupta, J. L. and Deheri, G. M. (1996). Effect of roughness on the behaviour of squeeze film in a spherical bearing. *Tribol. Trans.*, 39, 99–102.

12. Irmay, S. (1995). Flow of liquid through cracked media. *Bull. Res. Council Isr.*, 5A(1), 84.

13. Kuzma, D. C. (1964). Magnetohydrodynamic squeeze films. *J. Basic Engg. Trans. ASME*, 86, 441–444.

14. Kuzma, D. C., Maki, E. R., and Donnelly, R. J. (1964). The magnetohydrodynamic squeeze films. *J. Fluid Mec.*, 19, 395–400.

15. Majumdar. B. C. (1985). *Introduction to Tribology of Bearings*. Wheeler Co. Ltd., India: Wheeler Publisher.

16. Patel, K. C. and Gupta, J. L. (1979). Behavior of hydromagnetic squeeze film between porous plates. *Wear*, 56, 327–339.

17. Patel, K. C. and Hingu, J. V. (1978). Hydromagnetic squeeze film behavior in porous circular disks. *Wear*, 49, 239–246.

18. Patel, R. M. and Deheri, G. M. (2007). Magnetic fluid based squeeze film between porous conical plates. *Indus. Lub. Tribol.*, 59(3), 143–147.

19. Patel, R. M., Deheri, G. M., and Vahder, P. A. (2015). Hydromagnetic rough spongy circular step bearing. *Eastern Acad. J.*, 3, 71–87.

20. Prajapati, B. L. (1995). On Certain Theoretical Studies in Hydrodynamic and Electromagnetohydrodynamic Lubrication. Ph.D. Thesis, Sardar Patel University, VallabhVidyanagar, Gujarat, India.

21. Prakash, J. and Vij, S. K. (1973). Load capacity and time height relations for squeeze film between porous plates. *Wear*, 24, 309–322.

22. Shukla, J. B. (1965). Hydromagnetic theory of squeeze films. *ASME*, 87, 142–144.

23. Shukla, J. B. and Prasad, R. (1965). Hydromagnetic squeeze films between two conducting surfaces. *J. Basic Engg Trans. ASME*, 87, 818–823.

24. Sinha, P. C. and Gupta, J. L. (1974). Hydromagnetic squeeze films between porous annular disks. *J. Math. Phy. Sci.*, 8, 413–422.

25. Ting, L. L. (1975). Engagement behaviour of lubricated porous annular disks Part I: Squeeze film phase, surface roughness and elastic deformation effects. *Wear*, 34, 159–172.

26. Tonder, K. C. (1972). Surface distributed waviness and roughness. *First World Conf. Indus. Tribol.*, A3, 128.

27. Tzeng, S. T. and Saibel, E. (1967). Surface roughness effect on slider bearing lubrication. *J. Lub. Technol. Trans. ASME*, 10, 334–348.

28. Vadher, P. A., Deheri, G. M., and Patel, R. M. (2008). Hydromagnetic squeeze film between conducting porous transversely rough triangular plates. *Ann. Faculty Engg. Hunedoara*, 6, 155–168.

14 Primary and k – primary fuzzy ideals of Γ – semirings

Tilak Raj Sharma[a] and Ritu Sharma[b]

Department of Mathematics, Himachal Pradesh University Regional Centre, Khaniyara, Dharamshala, Himachal Pradesh-176218, India

Abstract

In this study, we summarize the prime, primary, and k–primary fuzzy ideals to Γ–semirings hypothetical consequences and characterise the primary fuzzy ideals of a Γ–semiring R to the primary ideals of R and the k–primary fuzzy ideals of a Γ–semiring R in terms of the k–primary ideals of R.

Keywords: Primary fuzzy ideals, k–primary fuzzy ideals, Γ–semiring

AMS Mathematics subject Classification (2020): 16Y60.

1. Introduction

By substituting ideals for elements Sharma et al. (2006) generalized the main ideals from commutative rings to non-commutative rings. Under the presumption that the ring is Noetherian, Stenstron (1975) provides this characterization of main ideals through their related primes. They were able to generalize some important findings for primary ideals in commutative rings to a non-commutative situation using the earlier method without having to use the Noetherian limitation. Several essential results concerning primary ideals and their radicals for non-commutative semirings. Zadeh (1965) introduced the concept of primary ideals for non-commutative semirings. L. A. Zadeh (1965) introduced the concept of fuzzy sets in 1965. Numerous mathematicians have used the idea of a fuzzy subset in the algebraic theory of groups and rings (Bhargavi and Eswarlal, 2015; Dutta, et al.; Jun and Lee, 1992). In algebra, the concept of Γ was first presented by N. Nobusawa in 1964. In a Γ – ring, the concept of fuzzy ideals was first articulated in 1992 by Jun and Lee. M. M. K. Rao first proposed the idea of Γ – semiring as a generalization of Γ – ring in 1995.

   The fact that fuzzy ideas proposed by Zadeh (1965) have been effectively applied to Γ – rings and semirings by Jun and Lee (1992) and that Γ – semiring is a generalization of semiring as well as Γ – ring serves as the inspiration for this research. The results of semirings from Joseph (2008) and Sharma (2015) characterized to primary fuzzy ideals of a Γ – semiring R in terms of primary ideals of R and k – primary fuzzy ideals of R in terms of k – primary ideals of R by starting with primary fuzzy ideals.

[a]trpangotra@gmail.com, [b]joyagnihotri@gmail.com

2. Preliminaries

Throughout this paper R represents a Γ–semiring. For preliminaries of Γ–semirings, we refer to the following definitions from Bhargavi and Eswarlal (2015), Dutta, et al., Rao (1995), and Sharma et al. (2006) which are necessary for this paper.

2.1 Definition
"Let S *and* Γ be non-empty sets. Then S is called a Γ – semigroup if there exists a mapping $S \times \Gamma \times S \to S$ denoted by $(x, \alpha, y) \to x\alpha y$ satisfying the condition $x\alpha(y\beta z) = (x\alpha y)\beta z$ for all $x, y, z \in S$ and for all $\alpha, \beta \in \Gamma$."

2.2 Definition
"Let $(R, +)$ and $(\Gamma, +)$ be two commutative semigroups. Then R called a Γ– semiring if there exists a mapping $R \times \Gamma \times R \to R$ denoted by $x\alpha y$ for all $x, y \in R$ and $\alpha \in \Gamma$ satisfying the following condition:

(i) $x\alpha(y + z) = x\alpha y + x\alpha z$
(ii) $(y + z)\alpha x = y\alpha x + z\alpha x$
(iii) $x(\alpha + \beta)z = x\alpha z + x\beta z$
(iv) $x\alpha(y\beta z) = (x\alpha y)\beta z$, for all $x, y, z \in R$ and $\alpha, \beta \in \Gamma$."

2.3 Definition
"A Γ–semiring R is said to have zero element if $0\alpha x = 0 = x\alpha 0$ and $x + 0 = x = 0 + x$ for all $x \in R$ and $\alpha \in \Gamma$."

2.4 Definition
"A Γ–semiring R is said to have an identity element if for all $x \in R$ there exist $\alpha \in \Gamma$ such that $1\alpha x = x = x\alpha 1$."

2.5 Definition
"Let X be a non-empty set. A mapping $\mu : X \to [0, 1]$ is called fuzzy subset of X."

2.6 Definition
"Let X be a non-empty set and I be a nonempty index set and let $(\lambda_i)_{i \in I}$ be a family of fuzzy sets of X. Then $\cup_{i \in I} \lambda_i$ and the intersection $\cap_{i \in I} \lambda_i$ of the family $(\lambda_i)_{i \in I}$ is defined by $\cup_{i \in I} \lambda_i = \sup\{\lambda_i (x) \mid i \in I\}$ and $\cap_{i \in I} \lambda_i = \inf\{\lambda_i (x) \mid i \in I\}$ for all $\in X$. Here *sup* and *inf* denote supremum and infimum, respectively."

2.7 Definition
"Let R be a Γ– semiring and μ_1, μ_2 be fuzzy left ideal [right ideal, fuzzy ideal]. Then the sum $\mu_1 + \mu_2$, product $\mu_1 \Gamma \mu_2$ and composition $\mu_1 \circ \mu_2$ of μ_1 and μ_2 are given by

$$(\mu_1 + \mu_2)(x) = \begin{cases} \sup_{x=a+b}[\min\{(\mu_1(a), \mu_2(b)\}], a, b \in R \\ 0 \qquad if\ for\ any\ a, b \in R, x \neq a + b \end{cases}$$

$$(\mu_1 \Gamma \mu_2)(x)$$

$$= \begin{cases} \sup_{x=a\alpha b}[\min\{(\mu_1(a), \mu_2(b)\}], & a, b \in R, \quad \alpha \in \Gamma \\ 0 & \text{if for any } a, b \in R, \quad \text{for any } \alpha \in \Gamma, \quad x \neq a\alpha b \end{cases}$$

$$(\mu_1 \circ \mu_2)(x) = \begin{cases} \sup_{x=\sum_{i=1}^n a_i \alpha_i b_i}[\min_{1 \leq i \leq n}[\min\{(\mu_1(a_i), \mu_2(b_i)\}]], & \\ \qquad a_i, b_i \in R, \quad \alpha_i \in \Gamma & \text{''} \\ 0 & \text{otherwise} \end{cases}$$

2.8 Definition
"Let R be a Γ – semiring. A non-constant fuzzy set $\lambda : R \to [0, 1]$ is a fuzzy left (right) ideal if (i) $\lambda(x + y) \geq min(\lambda(x), \lambda(y))$ (ii) $\lambda(x\alpha y) \geq \lambda(y)$, $(\lambda(x\alpha y) \geq \lambda(x))$. The fuzzy set λ is said to be the fuzzy ideal of Γ – semiring R if it is both left and a right fuzzy ideal of R."

2.9 Remark
Throughout this paper, R will denote a Γ – semiring with zero element "0" and identity element "1" unless otherwise stated.

3. Primary fuzzy ideals of a Γ-semiring

In a Γ–semiring, we define the fuzzy analogue of primary ideals as follows:

3.1 Definition
Let P be a fuzzy ideal of a Γ–semiring R. If either $P = \chi_R$ or P is nonconstant then P is primary fuzzy ideal and for any two fuzzy ideals λ and μ of R, $\lambda\Gamma\mu \subseteq P$ implies that either $\lambda \subseteq P$ or $\mu \subseteq r(P)$.

3.2 Definition
"Let λ be a fuzzy ideal of a Γ–semiring R. The fuzzy radical of λ, denoted by $r(\lambda)$, is defined by $r(\lambda) = \cap \{P|P \in \rho_\lambda\}$, *where* ρ_λ denote the family of all prime fuzzy ideals P of R such that $\lambda \subseteq P$ *and* $\lambda_* \subseteq P_*$, where $\lambda_* = \{x \in R| \lambda(x) = \lambda(0)\}$ and $P_* = \{x \in R|P(x) = P(0)\}$."

Note

(i) $\rho_\lambda \neq \phi$ *as* $\chi_R \in \rho_\lambda$.
(ii) If $\lambda(0) = 1$, then every prime fuzzy ideal P contains λ is in ρ_λ.

The following theorem is proved by Sharma and Sharma (2023).

Theorem 3.3

"Let P be a non-constant fuzzy ideal of a Γ–semiring R. Then P is prime if and only if there exist a prime ideal π of R $(\pi \neq R)$ such that $P_r \in \{\pi, R\}$, for all $r \in [0, 1]$."

In terms of primary fuzzy ideals of a Γ – semiring R, the following theorem defines primary ideals of R.

Theorem 3.4

Let P be a non-constant fuzzy ideal of a Γ – semiring R. Then P is a primary fuzzy ideal of R if and only if $P_r \in \{\pi, R\}$ for all $r \in [0, 1]$, where $\pi(\neq R)$ is a primary ideal of R.

Proof: Let the empty set ϕ and all the ideals of R make up the set $I(R)$. The decreasing function $\Psi : [0, 1] \to I(R)$ given by $\Psi(r) = P_r$ takes on at least one non-empty set aside from R, since P is non-constant. In fact, we show that for all $r \in [0, 1]$, where $t = P(0) > u = P(1)$, it takes precisely two values $P_t \neq \phi$ and $P_u = R$. If not, then there exists $s \in [0, 1]$, with $t > s$ such that $P_t \subsetneq P_s \subsetneq R$. Define two fuzzy ideals λ and μ of R by $\lambda_r = \begin{cases} P_s & if\ r \in (0,1] \\ R & if\ r = 0 \end{cases}$ and $\mu_r = \begin{cases} P_r & if\ r > s \\ R & if\ r \leq s \end{cases}$. As in Theorem 3.3 λ $\Gamma \mu \subseteq P$ and λ is not contained in P, since $\lambda_r = P_s \not\subseteq P_t$. Here, we demonstrate how the statement "P is primary" will be contradicted by $\mu \not\subseteq r(P)$. Since $P_s \neq R$, there exists $m, 0 < m < s$, such that $P_m \neq R$. For, if $P_m = R$ for all $0 < m < s$, then for any $x \in R$, we have $P(x) = sup\{i \in [0, 1] \mid x \in P_i\} \geq sup\{m \in [0, 1] \mid 0 < m < s\} = s$ implying that $x \in P_s$. This contradicts the fact that $P_s \neq R$. Moreover $m < s$ implies $P_s \subseteq P_m$, so that $P_m \neq \phi$ as $P_s \neq \phi$. Since $\phi \subsetneq P_m \subsetneq R$ and P_m is an ideal of R, therefore there exists a prime ideal $I (\neq R)$ of R such that $P_m \subseteq I$.

Define a fuzzy ideal P' of R by $P_r' = \begin{cases} I & if\ r > m \\ R & if\ r \leq m \end{cases}$. Then by Theorem 3.3, P' is a prime fuzzy ideal of R. $r > m \Rightarrow P_r \subseteq P_m \subseteq I = P_r'$, and $r \leq m \Rightarrow P_r' = R$, so $P_r \subseteq P_r'$. Thus $P \subseteq P'$. Also $P_m \neq \phi$, so $P_* \subseteq P_m \subseteq I = P_{*}'$. Further $P'_s = I \neq R$. Now $P' \supseteq P, P'_*$. $\supseteq P_*, P'_s \neq R$ and for any family $(\lambda_i)_{i \in I}$ of fuzzy ideals of R, $(\cap \lambda_i)_r = \cap (\lambda_i)_r, r \in [0, 1]$, thus $(r(P))_s \neq R$. However $\mu_s = R$. Therefore $\mu_s \not\subseteq (r(P))_s$ and consecutively $\mu \not\subseteq r(P)$ Thus there exist an ideal $\pi(\phi \subsetneq \pi \subsetneq R)$ with $P_r \in \{\pi, R\}$ for $r \in [0, 1]$. Now, to prove π is primary. Let $\lambda \Gamma \mu \subseteq \pi$, where μ and λ be two ideals of R and the characteristic function χ_μ and χ_λ satisfy $\chi_\mu \Gamma \chi_\lambda \subseteq \chi_\lambda \Gamma_\mu \subseteq \chi_\pi \subseteq P$, because

$$(\chi_\pi)_r = \begin{cases} \pi & if\ 0 < r < 1 \\ R & if\ r = 0 \end{cases} \tag{1}$$

Thus, either $\chi_\mu \subseteq P$ or $\chi_\lambda \subseteq r(P)$ If $\chi_\mu \subseteq P$, then $\mu \subseteq \pi$. Suppose $\chi_\mu \not\subseteq P$ and $\chi_\lambda \subseteq r(P)$. Let π' be any prime ideal of R containing π. Now, $P_r = \begin{cases} \pi & if\ r > n \\ R & if\ r \leq n \end{cases}$, for some $n \in [0, 1)$, since $P_r \in \{\pi, R\}$, for all $r \in [0, 1]$. Define

$$P_r' = \begin{cases} \pi' & if\ r > n \\ R & if\ r \leq n \end{cases}. \tag{2}$$

Since $\pi \subseteq \pi'$, so $P' \supseteq P$ and $P_*' \supseteq P_*$. Thus, $\chi_\lambda \subseteq r(P) \Rightarrow \chi_\lambda \subseteq P'$. This gives $\lambda = (\chi_\lambda)_I \subseteq (P')_I = \pi'$ and so $\lambda \subseteq r(\pi)$. Hence, π is primary. Conversely, assume that that $P_r \in \{\pi, R\}$, π is primary ideal. If possible, let P is not primary. Then there exits fuzzy ideals σ and τ with $\sigma \Gamma \tau \subseteq P$, but $\sigma \not\subseteq P$ and $\tau \not\subseteq r(P)$. $\sigma_s \not\subseteq P_s$ and $\tau_t \not\subseteq (r(P))_t$, for some $s, t \in [0, 1]$. $\sigma_s \not\subseteq P_s \Rightarrow P_S = \pi$. Further, $\tau_t \not\subseteq (r(P))_t$ with $P' \supseteq P$ and $P_*' \supseteq P^* = \pi$ such that $\tau_t \not\subseteq P_t'$. So $P_t' = P_*'$. Consequently, $\tau_t \not\subseteq r(\pi)$. Let $m = min\ (s, t)$ Then $s \geq m$ and $t \geq m$ implies that $\sigma_m \subseteq \pi$ and $\tau_m \not\subseteq \sqrt{\pi}$. But $\sigma \Gamma \tau \subseteq \pi$ implies that $P(a \alpha b) \geq min\ (\sigma(a), \tau(b))$. Thus, for any $z = \Sigma_i x_i \alpha_i y_i$ in $\sigma_m \Gamma \tau_m$, where x_i, gives $\sigma_m \Gamma \tau_m \subseteq \pi$, a contradiction. Hence P is primary.

We will now derive the fuzzy equivalents of several outcomes using the characterization theorem 3.4.

Theorem 3.5

(i)   Let $P_1, P_2, P_3, \ldots, P_n$ be primary fuzzy ideals of R such that $r(P_i) = \mu$ ($i = 1, 2, 3, \ldots, n$) Then $P = \cap_{i=1}^n P_i$ is primary and $r(P) = \mu$.

(ii) Let R and R' be two Γ–semirings and $T: R \rightarrow R'$ be onto homomorphism. Let $K = \{x \in R \mid x = a + b, T(a) = T(b)\} \subseteq \lambda_*$ and λ a fuzzy ideal of R such that both λ_* and $r(\lambda_*)$ are k–ideals. If λ is primary, then $T(\lambda)$ is primary. Additionally, if the range λ is finite and $T(\lambda)$ is primary then λ is primary.

Proof:

(i)   The result follows by the characterization of primary fuzzy ideals $P_1, P_2, P_3, \ldots, P_n$ and (cf. [8], theorem 5.2.13 (i)(a)). By Theorem 4.4, P is primary. Furthermore, using (cf. [8], theorem 5.2.2(iv)) and the fact $(\cap_{i=1}^n P_i)_s = \cap_{i=1}^n (P_i)_s$ and $(r(\cap_{i=1}^n P_i))_s = r(\cap_{i=1}^n P_i)_s$, we have $r(P) = r(\cap_{i=1}^n P_i) = (\cap_{i=1}^n (P_i) = \mu$

(ii) We first show that $(T(\lambda))_r = T(\lambda_r)$ for all $r \in [0, 1]$ if range λ is finite. For this, let $y \in T(\lambda_r)$. Then there exists $x \in \lambda_r$ such that $T(x) = y$. Now, $x \in \lambda_r$ implies $\lambda(x) \geq r$ and therefore $T(\lambda)(y) = {}^{sup}_{T(x)=y}\{\lambda(x)\} \geq r$. Thus $y \in (T(\lambda))_r$. Hence $T(\lambda_r) \subseteq (T(\lambda))_r$. To show that $(T(\lambda))_r \subseteq T(\lambda_r)$, let $y \in (T(\lambda))_r$ then $(T(\lambda))(y) \geq r$, that is ${}^{sup}_{T(z) = y}\{\lambda(z)\} \geq r$. Since range of λ is finite, therefore, ${}^{sup}_{x \in E} \lambda(x) = \lambda(x_0)$, where E is any subset of R and $x_0 \in E$. Thus there exists $z_0 \in R$ such that $T(z_0) = y$ and $\lambda(z_0) = {}^{sup}_{T(z)=y}\{\lambda(z)\} \geq r$. Hence $y \in T(\lambda_r)$ and therefore $T(\lambda_r) = (T(\lambda))_r$. Now, by theorem 3.4, $\lambda r = \begin{cases} \pi & if \, r > m \\ R & if \, r \leq m \end{cases}$, where π is a primary ideal of R and $m = \sup \{i \in [0, 1] \mid \lambda_i = R\}$. Thus, $(T(\lambda))_r = T(\lambda r) = \begin{cases} T(\pi) & if \, r > m \\ R' & if \, r \leq m \end{cases}$. Since $\lambda_* = \pi \supseteq K$, by assumption both π and $r(\pi)$ are k–ideals. Thus, it follows from (cf. [8], theorem 5.2.15) that $T(\pi)$ is primary and conversely, if $T(\pi)$ is primary then π is primary.

3.6 Definition

A fuzzy ideal λ of R is said to be a maximal fuzzy ideal if (i) λ is not constant. (ii) $\lambda \subseteq \mu$ then either $\lambda_* = \mu_*$ or $\mu_* = \chi_R$, where $\lambda_* = \{x \in R \mid \lambda(x) = \lambda(0)\}$, $\mu_* = \{x \in R \mid \mu(x) = \mu(0)\}$.

Theorem 3.7

Let P be a non-constant fuzzy ideal of R. Then there exists a maximal ideal π of R such that $P_r \in \{\pi, R\}$, for all $r \in [0, 1]$ if and only if P is a maximal fuzzy ideal of R.

k-primary fuzzy ideals of a Γ–semiring

4.1 Definition

Let λ be a fuzzy ideal of a Γ – semiring R. The fuzzy k – radical of λ, denoted by $k - r(\lambda)$ and is defined as $k - r(\lambda) = \cap_i \{\pi_i \mid \pi_i \in \wp_s(\lambda)\}$, $\wp_s(\lambda)$ is the set of all k–prime fuzzy ideals π_i of R such that $\lambda \subseteq \pi_i$ and $\lambda_* \subseteq (\pi_i)_*$, where $\lambda_* = \{x \in R \mid \lambda(x) = \lambda(0)\}$ and $(\pi_i)_* = \{x \in R \mid \pi_i(x) = \pi_i(0)\}$.

4.2 Definition

Let λ be a fuzzy ideal of R. Then λ is a k – primary fuzzy ideal if for any two fuzzy ideals μ and τ, $\mu\,\tau \subseteq \lambda \Rightarrow$ either $\mu \subseteq \lambda$ or $\tau \subseteq (k - r(\lambda))$.

Theorem 4.3

Let λ be a fuzzy ideal of R. Then λ is a k – primary fuzzy ideal of R if there exists a k–primary ideal P of R such that $f_\lambda(i) = \begin{cases} P & if\ i > d \\ R & if\ i \le d \end{cases}$, where $d = \sup\{i \in [0, 1]$ $\mid f_\lambda(i) = R\}$.

Proof: Let $I(R) = ideal(R) \cup \{\phi, R\}$. Clearly, the decreasing function $f_\lambda: [0, 1] \to I(R)$ given by $f_\lambda(i) = \lambda_r$ takes on at least one element of $ideal(R)$. In fact it takes on exactly two values of $r \in [0, 1]$, that is $f_\lambda(t) \ne \phi$ and $f_\lambda(d) = R$ where $t = \lambda(0) = \sup\{i \in [0, 1] \mid f_\lambda(i) = R\}$. If not, then there exists $s \in [0, 1]$ with $t > s$ such that $\phi \ne f_\lambda(t) \subsetneq f_\lambda(s) \subsetneq R$. Define two fuzzy ideals π and μ as

$$f_\pi(r) = \begin{cases} f_\lambda(s) & if\ r > (0,1] \\ R & r = 0 \end{cases} \quad \text{and} \quad f_\mu(r) = \begin{cases} f_\lambda(r) & if\ r > s \\ R & r \le s \end{cases}.$$

Therefore, $\pi\,\Gamma\,\mu \subseteq \lambda$ and $\pi \not\subseteq \lambda$. For contradiction, it only remains to show that $\mu \not\subseteq k - r(\lambda$. Since $f_\lambda(s) \ne R$, there exists m, $0 < m < s$ such that $f_\lambda(m) \ne R$. If $f_\lambda(m) = R$, $0 < m < s$ then for any $x \in R$ we have $\lambda(x) = \sup\{i \in [0, 1] \mid x \in f_\lambda(i)\} \ge \sup\{m \in [0, 1] \mid 0 < m < s\} = s$. This implies that $x \in f_\lambda(s)$, contradicting the fact that $f_\lambda(s) \ne R$. Further $m < s \Rightarrow f_\lambda(s) \subseteq f_\lambda(m)$, so we get $f_\lambda(m) \ne \phi$ as $f_\lambda(m) \ne \phi$. Since $\phi \subsetneq f_\lambda(m) \subsetneq R$, it will be contained in a maximal ideal M of R which is a k–prime (cf.[5] theorem 2.2.1, 3.2.13 and cf. [8], corollary 5.1.6). Hence, $f_\lambda(m) \subseteq M$. Define a fuzzy ideal τ of R by $f_\tau(r) = \begin{cases} M & if\ r > m \\ R & if\ r \le m \end{cases}$. Now the characterization of k- prime fuzzy ideals implies that τ is a k – prime fuzzy ideal of R. Moreover, $r > m \Rightarrow$ that $f_\lambda(r) \subseteq f_\lambda(m) \subseteq M = f_\tau(r)$ and $r \le m \Rightarrow f_\tau(r)$, $= R$ so that $f_\lambda(r) \subseteq f_\tau(r)$. Thus $\lambda \subseteq \tau$. Again, $f_\tau(s) = M \ne R$, since $s > m$. Now the facts that τ is k – prime fuzzy ideal of R, $\lambda \subseteq \tau$, $\lambda_* \subseteq \tau_*$ that $f_{k-r(\lambda)}(s) \subseteq f_\tau(s)$. But $f_\tau(s) \ne R$. Therefore, $f_\mu(s) \not\subseteq f_{k-r(\lambda)}(s)$ and consequently $\mu \subseteq (k - r(\lambda))$. Thus there exists an ideal P of R with $f_\lambda(r) = \begin{cases} P & if\ r > d \\ R & if\ r \le d \end{cases}$. Let A and B two ideals of R such that $A\Gamma B \subseteq P$. Then the characteristic functions χ_A and χ_B satisfy $\chi_A\,\Gamma\,\chi_B \subseteq \chi_{A\Gamma B} \subseteq \chi_P \subseteq \lambda$, because $f_{\chi_P}(r) = \begin{cases} P & if\ r \in (0,1] \\ R & if\ r = 0 \end{cases}$. k–primary character of λ have implies that either $\chi_A \subseteq \lambda$ or $\chi_B \subseteq k$–$r(\lambda)$. If $\chi_A \subseteq \lambda$ then $A \subseteq P$. Suppose that $\chi_A \subseteq \lambda$. Then $\chi_B \subseteq k$–$r(\lambda)$. Let P' be any k–prime ideal of R containing P (cf., [5] theorem 2.2.1, 3.2.13 and cf. [8], corollary 5.1.6). Since $f_\lambda(r) = \begin{cases} P & if\ r > d \\ R & if\ r \le d \end{cases}$, we have a k – prime fuzzy ideal λ' of R defined by $f_{\lambda'}(r) = \begin{cases} P' & if\ r > d \\ R & if\ r \le d \end{cases}$, containing λ and $\lambda'_* \supseteq \lambda_*$. Hence $\chi_B \subseteq k - r(\lambda) \subseteq \lambda'$ which gives $B \subseteq f_{\chi B}(1) \subseteq f_{\lambda'}(1) = P'$ So, $B \subseteq k - r(P)$. Conversely, let $f_\lambda(r) = \begin{cases} P & if\ r > d \\ R & if\ r \le d \end{cases}$, where P is a k–primary ideal of R.

Suppose that λ is not k–primary. Then $\pi\Gamma\tau\subseteq\lambda$ gives that $\mu\not\subseteq\lambda$ and $\tau\not\subseteq k\text{-r}(\lambda)$. This implies that $f_\mu(s)\not\subseteq f_\lambda(s)$ and $f_\tau(t)\not\subseteq f_{k-r(\lambda)}(t)$ for some $s, t\in[0,1]$. So $f_\mu(s)\not\subseteq f_\lambda(s)$ implies that $f_\lambda(s)=P$, since every ideal is contained in R. Further, $f_\tau(t)\not\subseteq f_{k-r(\lambda)}(t)$ implies that there exists a k–prime fuzzy ideal λ' of R with $\lambda'\supseteq\lambda$ and $\lambda'_*\supseteq\lambda_*$ $=P$ such that $f_\tau(t)\not\subseteq f_{\lambda'}(t)$. This implies that $f_{\lambda'}(t)=\lambda'_*$. Thus there exists k–prime ideal λ'_* of R containing P such that $f_\tau(t)\not\subseteq\lambda'_*$. Consequently, $f_\tau(t)\not\subseteq k\text{-r}(P)$. Let d $=min(s,t)$. Then $s\geq d$ and $t\geq d$ implying that $f_\mu(k)\subseteq P$ and $f_\tau(d)\not\subseteq k\text{-r}(P)$. But $\mu\Gamma\tau$ $\subseteq\lambda\Rightarrow$ that $\lambda(a\alpha b)\geq min(\mu(a),\tau(b))$, $\alpha\in\Gamma$. Thus, for any $z=\Sigma x_i\,\alpha_i\,y_i$ in $f_\mu(d)\,\Gamma$ $f_\tau(d)$ where $x_i\in f_\mu(d), y_i\in f_\tau(d)$ and $\alpha_i\in\Gamma$, we have $\lambda(z)\geq d$, since $\mu(x_i)\geq d$ and $\tau(y_i)$ $\geq d$. Hence, $z\in f_\lambda(d)=P$ will give $f_\mu(d)\,\Gamma\,f_\tau(d)\subseteq P$, contradicting the k–primary character of P. So λ is a k–primary fuzzy ideal of R.

Corollary 4.4

Let $\lambda_1,\lambda_2,\lambda_3,\ldots\lambda_n$ be k–primary fuzzy ideals of a Γ–semirings R such that $k-r(\lambda_i)$ $=\mu$. Then $\lambda=\cap_{i=1}^n\lambda_i$ is k–primary and $k-r(\lambda)=\mu$.

Conclusion/Novelty

The main objective of this paper is to derive a relationship between the primary fuzzy ideals of a Γ–semiring R to the primary ideals of R and k – primary fuzzy ideals in terms of k – primary ideals of R. It recommended that the concepts presented in this article have a lot of potential for nourishing and one can explore the same in fuzzy semiprime, fuzzy maximal and fuzzy irreducible ideals for Γ–semirings.

References

1.   Bhargavi, Y. and Eswarlal, T. (2015). Fuzzy gamma semirings. *Int. J. Pure Appl. Math.*, 98, 339–349.
2.   Dutta, T. K., Sardar, S. K., and Goswami, S. Operations on fuzzy ideals of Γ–semirings. *Proc. Nat. Sem. Alg. Anal. Dis. Math.*, arXiv: 1101. 4791v1 [math. RA].
3.   Jun, Y. B. and Lee, C. Y. (1992). Fuzzy Γ–Rings. *Pusom Young man Math J.*, 8, 63–70.
4.   Joseph, R. (2008). Ph.D. thesis. Some problems on group graded semirings and their smash products.
5.   Liu, W. J. (1982). Fuzzy invariant subgroups and fuzzy sets and systems. 8, 133–139.
6.   Nobusawa, N. (1964). On a generalization of the ring theory. *Osaka J. Math.*, 181–189.
7.   Rao, M. M. K. (1995). Γ–Semiring I. *Southeast Asian Bull. Math.*, 19, 281–287.
8.   Shweta, G. (2018). Ph.D. thesis. On some problems in a Γ–semiring and their ideals.
9.   Sharma, R. P., Gupta, J. R., and Ranju, B. (2006). Primary ideals and fuzzy ideals in a ring. *South Asian Bull. Math.*, 30, 731–744.
10. Sharma, T. R. (2015). Primary and G-primary fuzzy ideals of a semiring. *Himachal Pradesh University J.*, 03, 223–228.
11. Sharma, R. P., Sharma, T. R. and Joseph, R. (2011). Primary ideals in non-commutative semirings. *J. Southeast Asian Bull. Math.*, 35, 345–360.
12. Sharma, T. R. and Sharma, R. (2023). Prime fuzzy ideals of a Γ–semiring-1. *Indian J. Sci. Technol.* 16, 2441–2446.
13. Stenstron, Bo. (1975). Rings of quotients (Grundlehren Math. Wiss. Bd 217), New York: Springer-Verlag.
14. Zadeh, L. A. (1965). Fuzzy sets. *Inform. Control*, 8, 330–353.

15 Some bounds of the neighborhood version of Banhatti-Sombor index and modified Sombor index

Shivani Rai[1,a] and Biswajit Deb[2]

[a]Assistant Professor, Research Scholar, Department of Mathematics, Sikkim Manipal Institute of Technology, the ICFAI University, Sikkim, India.

[b]Associate Professor, Department of Mathematics, Sikkim Manipal Institute of Technology, India

Abstract

Numerical parameters that are highly correlated with the branching pattern of chemical compounds are known as topological indices. Apart from investigating their chemical applications, exploring their mathematical properties has caught the attention of many researchers. In this paper, we introduce the neighborhood variants of the Banhatti-Sombor index and the modified Sombor index. Furthermore, their bounds are explored in terms of graph parameters like the order of $\Gamma(n(\Gamma))$, size of $\Gamma(m(\Gamma))$, maximum degree of $\Gamma(\Delta(\Gamma))$ and some neighborhood degree-based in- dices like the neighborhood modified Zagreb index $(^{nm}M_2)$, neighborhood Harmonic index (NH), fifth NDe index (ND_5), neighborhood forgotten index (F^*), neighbor-hood Sombor index (NSO), neighborhood inverse sum index (NI) and fourth NDe index (ND_4). Additionally, extremal graphs have been characterized with respect to these indices.

Keywords: Molecular graph, topological index, neighborhood degree, neighborhood Banhatti Sombor index, neighborhood modified Sombor index

AMS Classification

MSC (2010): Primary: 05C35; Secondary: 05C07, 05C40

1. Introduction

Throughout this article simple connected graphs are considered, unless otherwise stated. One can refer to Harary (1969) for basic terminologies and concepts on graphs. A molecular graph (Γ) is a graphical depiction of a chemical compound where the vertices correspond to the atoms and the edges correspond to the bonds between those atoms. A numerical parameter that is mathematically deduced from a molecular graph is known as a topological index. Topological indices have experienced a surge in their use over the past eight decades ever since their inception (Weiner 1947; Gutman 2013; Redžepović and Furtula 2020; Adnan et al., 2022). The physico-chemical properties, biological activities of chemical

[a]shivanirai866@gmail.com

Table 15.1 Some well-known *NDSBI*

Name	Notation	Formula
Neighborhood modified Zagreb index	nmM_2	$\sum_{kl \in E(\Gamma)} \frac{1}{\delta(k)\delta(l)}$
Fourth NDe index	ND_4	$\sum_{kl \in E(\Gamma)} \frac{1}{\sqrt{\delta(k)\delta(l)}}$
Neighborhood Harmonic index	NH	$\sum_{kl \in E(\Gamma)} \frac{2}{\delta(k)+\delta(l)}$
Neighborhood inverse sum index	NI	$\sum_{kl \in E(\Gamma)} \frac{\delta(k)\delta(l)}{\delta(k)+\delta(l)}$
Fifth NDe index	ND_5	$\sum_{kl \in E(\Gamma)} \frac{(\delta(k))^2+(\delta(l))^2}{\delta(k)\delta(l)}$
Neighborhood Forgotten index	F^*_N	$\sum_{kl \in E(\Gamma)} (\delta(k))^2 + (\delta(l))^2$
Fifth Geometric-arithmetic index	GA_5	$\sum_{kl \in E(\Gamma)} \left(\frac{\sqrt{2\delta(k)\delta(l)}}{\delta(k)+\delta(l)} \right)$
Neighborhood Sombor index	NSO	$\sum_{kl \in E(\Gamma)} \sqrt{(\delta(k))^2 + (\delta(l))^2}$

compounds, medicines, and nanomaterials have been predicted using different indices up to this point some of which can be seen in (de Julián-Ortiz et al. 1998; Nadeem, Zafar, and Zahid 2016; Kwun et al., 2017; Mondal et al., 2021). These are based on the neighborhood degree-sum of a vertex l, i.e., the sum of the degrees of neighbors of a vertex l, denoted by $\delta(l)$. Some neighborhood degree sum-based indices (*NDSBI*) are listed in Table 15.1.

The introduction of a novel class of degree-based topological indices known as the Sombor indices has attracted the interest of researchers recently (Aguilar-Sánchez et al., 2021; Réti, Döslic, and Ali, 2021). In 2021 Sombor index and its two other variants were defined Gutman (2021).

In Cruz, Gutman, and Rada (2021), the Sombor index of chemical graphs was studied and extremal graphs were characterized. Some chemical applications of this index were explored in (Deng, Tang, and Wu, 2021; Redžepović 2021). Later, Kulli introduced the first Banhatti-Sombor index, the first reduced Banhatti-Sombor index, first μ-Banhatti-Sombor index in Kulli (2021b). In the same work, the exact formulas of these indices for some nanostructures were also studied. Other variants of the modified Sombor index, were also introduced and exact expressions for certain chemical structures were deduced. In Huang and Liu (2021), some bounds of the modified Sombor index as well as for the modified spectral radius and energy were determined. For some interesting results based on the modified Sombor matrix, one can refer to Zuo et al. (2022). Subsequently, the neighborhood version of the Sombor index $NSO(\Gamma)$ was proposed known as the neighborhood Sombor index Kulli (2021a). Additionally, the second, third and fourth Sombor indices were also introduced in the same paper and all these indices were computed for some nanostructures.

Thus, motivated by the above results, in this work, the neighborhood version of the Banhatti Sombor index and modified Banhatti index are defined as

$$NBSO_1(\Gamma) = \sum_{kl \in E(\Gamma)} \sqrt{\frac{1}{(\delta(k))^2} + \frac{1}{(\delta(l))^2}}.$$

and

$$^mNSO_1(\Gamma) = \sum_{kl \in E(\Gamma)} \frac{1}{\sqrt{(\delta(k))^2 + (\delta(l))^2}}.$$

Additionally, some bounds of these indices are determined in terms of some graph parameters like $n(\Gamma)$, $m(\Gamma)$ and $\Delta(\Gamma)$, and other neighborhood degree-based topological indices. Furthermore, extremal graphs with respect to these indices are characterized.

2. Preliminaries

Lemma 2.1. *Lin et al. (2021) For any two positive real numbers x and y, we have*

$$\frac{2\sqrt{2}(x^2 + y^2 + xy)}{3(x + y)} \leq \sqrt{x^2 + y^2} \leq \frac{\sqrt{2}(x^2 + y^2)}{x + y}.$$

Here, equality is achieved iff x = y.

Lemma 2.2. *Lin et al. (2021) If $x_i > 0$, $y_i > 0$, $z > 0$ and $\{i \in N \mid 1 \leq i \leq n\}$, then the inequality is obtained as follows:*

$$\sum_{i=1}^{n} \frac{x_i^{z+1}}{y_i^z} \geq \frac{\left(\sum_{i=1}^{n} x_i\right)^{z+1}}{\left(\sum_{i=1}^{n} y_i\right)^z}.$$

Here, equality is achieved iff $\frac{x_1}{y_1} = \frac{x_2}{y_2} \cdots = \frac{x_n}{y_n}$.

Lemma 2.3. *Dragomir (1984) Let $x = (x_1, x_2, \ldots, x_n)$, $y = (y_1, y_2, \ldots, y_n)$ be sequences of real numbers and $w = w_1, w_2, \ldots, w_n$, $z = z_1, z_2, \ldots, z_n$ be non-negative, then*

$$\sum_{i=1}^{n} z_i \sum_{i=1}^{n} w_i x_i^2 + \sum_{i=1}^{n} w_i \sum_{i=1}^{n} z_i y_i^2 \geq 2 \sum_{i=1}^{n} w_i x_i \sum_{i=1}^{n} z_i y_i.$$

Here, equality is achieved iff $x = y = (c, c, \ldots, c)$ is a constant sequence for $w_i > 0$ and z_i and $\{i \in N \mid 1 \leq i \leq n\}$.

3. Main results

3.1 Bounds of neighborhood Banhatti Sombor index
Theorem 3.1. *Let Γ denote a graph with $n \geq 3$. Then,*

$$\sqrt{2}ND_4 \leq NBSO(\Gamma) \leq \sqrt{2}(n-1)\Delta \ ^{nm}M_2(\Gamma).$$

The left equality is achieved iff $\delta(k) = \delta(l)$ for every $kl \in E(\Gamma)$ and the right equality is achieved if $\Gamma = K_n$.

Proof. By arithmetic-geometric mean inequality,

$$NBSO(\Gamma) = \sum_{kl \in E(\Gamma)} \sqrt{\frac{1}{\delta(k)^2} + \frac{1}{\delta(l)^2}} \geq \sum_{kl \in E(\Gamma)} \sqrt{\frac{2}{\delta(k)\delta(l)}} = \sqrt{2} \sum_{kl \in E(\Gamma)} \sqrt{\frac{1}{\delta(k)\delta(l)}} \quad (1)$$

$$= \sqrt{2}ND_4.$$

Since arithmetic and geometric mean are equal iff all the terms of the series are equal, so the left equality is achieved iff $\delta(k) = \delta(l)$, for every $kl \in E(\Gamma)$. Now, for the upper bound, we need to show $NBSO(\Gamma) \leq \sqrt{2}(n-1)\Delta {}^{nm}M_2(\Gamma)$.

It is clear that when $\delta(k) = \delta(l) = (n-1)\Delta$, for every $kl \in E(\Gamma)$ i.e., if $\Gamma = K_n$. equality is obtained.

Theorem 3.2. *Let Γ denote a graph with $n \geq 3$. Then,*

$$NBSO(\Gamma) \geq \sqrt{2}NH(\Gamma). \quad (2)$$

Here, equality is achieved iff $\delta(k) = \delta(l)$ for every $kl \in E(\Gamma)$.

Proof. Applying the left inequality in Lemma 2.1 and by taking $x = \frac{1}{\delta(k)}$ and $y = \frac{1}{\delta(l)}$, we have

$$NBSO(\Gamma) = \sum_{kl \in E(\Gamma)} \sqrt{\frac{1}{\delta(k)^2} + \frac{1}{\delta(l)^2}} \geq \sum_{kl \in E(\Gamma)} \frac{2\sqrt{2}\left(\frac{\delta(l)}{\delta(k)} + \frac{\delta(k)}{\delta(l)} + 1\right)}{3(\delta(k) + \delta(l))}$$

$$= \sum_{kl \in E(\Gamma)} \frac{2\sqrt{2}(1+1+1)}{3(\delta(k) + \delta(l))}$$

which gives inequality (2). From Lemma 2.1, equality is achieved iff $\delta(k) = \delta(l)$ for all $kl \in E(\Gamma)$.

Theorem 3.3. *Let Γ denote a graph, $n \geq 3$. Then,*

$$NBSO(\Gamma) \leq \frac{1}{2\sqrt{2}}ND_5(\Gamma). \quad (3)$$

Here, equality is achieved if Γ is a disjoint union of P_3.

Proof. Using the right inequality in Lemma 2.1 we have,

$$NBSO(\Gamma) = \sum_{kl \in E(\Gamma)} \sqrt{\frac{1}{\delta(k)^2} + \frac{1}{\delta(l)^2}} \leq \sum_{kl \in E(\Gamma)} \sqrt{2}\left(\frac{\delta(l)}{\delta(k)} + \frac{\delta(k)}{\delta(l)}\right)\frac{1}{2+2}$$

$$= \sum_{kl \in E(\Gamma)} \sqrt{2}\left(\frac{\delta(l)}{\delta(k)} + \frac{\delta(k)}{\delta(l)}\right)\frac{1}{4}.$$

which gives right inequality in (3), as the minimum value of $\delta(k)$ for any $k \in V$ (Γ) is 2. From Lemma 2.1 equality is achieved iff $\delta(k) = \delta(l)$, for all $kl \in E(\Gamma)$. If Γ is a disjoint union of P_3 then, $\delta(k) = \delta(l) = 2$, for all kl $E(\Gamma)$. Therefore, the upper bound is achieved.

Theorem 3.4. *Let Γ denote a graph, $n \geq 3$. Then,*

$$NBSO(\Gamma) \geq \frac{\sqrt{2}}{3[(n-1)\Delta)]}ND_5 + \frac{\sqrt{2}}{3}NH(\Gamma). \quad (4)$$

Here, equality is achieved if $\Gamma = K_n$.
Proof. Applying Lemma 2.1,

$$NBSO(\Gamma) = \sum_{kl \in E(\Gamma)} \sqrt{\frac{1}{\delta(k)^2} + \frac{1}{\delta(l)^2}} \geq \sum_{kl \in E(\Gamma)} \frac{2\sqrt{2}\left(\frac{\delta(l)}{\delta(k)} + \frac{\delta(k)}{\delta(l)} + 1\right)}{3\left(\delta(k) + \delta(l)\right)}$$

$$\geq \frac{2\sqrt{2}}{3} \sum_{kl \in E(\Gamma)} \left(\frac{\frac{\delta(k)^2 + \delta(l)^2}{\delta(k)\delta(l)}}{(n-1)\Delta + (n-1)\Delta}\right) + \sum_{kl \in E(\Gamma)} \frac{2\sqrt{2}}{3(\delta(k) + \delta(l))}$$

which gives left inequality in (3). From Lemma 2.1, equality is obtained when $\delta(k)$ = $\delta(l)$, for all $kl \in E(\Gamma)$. If $\Gamma = K_n$ i.e., when $\delta(k) = \delta(l) = (n-1)\Delta$, for all $kl \in E(\Gamma)$, the lower bound is attained.

Theorem 3.5. *Let* Γ *denote a graph with* $n \geq 3$. *Then*

$$NBSO(\Gamma) \leq \sqrt{^{nm}M_2(\Gamma)ND_5}. \tag{5}$$

Here, equality is achieved iff $\sqrt{\delta(k)^2 + \delta(l)^2}$ *is a constant for every* $kl \in E(\Gamma)$.

Proof. Let $z = 1$, $z = 1$, $x_i = \sqrt{\frac{1}{\delta(k)^2} + \frac{1}{\delta(l)^2}}$ and $y_i = \frac{1}{\delta(k)\delta(l)}$ in Lemma 2.2 then we get,

$$\frac{\left(\sum_{kl \in E(\Gamma)} \sqrt{\frac{1}{\delta(k)^2} + \frac{1}{\delta(l)^2}}\right)^2}{\sum_{kl \in E(\Gamma)} \frac{1}{\delta(k)\delta(l)}} \leq \sum_{kl \in E(\Gamma)} \frac{\frac{1}{\delta(k)^2} + \frac{1}{\delta(l)^2}}{\frac{1}{\delta(k)\delta(l)}} = \sum_{kl \in E(\Gamma)} \left(\frac{\delta(l)}{\delta(k)} + \frac{\delta(k)}{\delta(l)}\right)$$

Therefore,

$$\frac{[NBSO(\Gamma)]^2}{^{nm}M_2(\Gamma)} \leq ND_5,$$

which gives us the required inequality (5). Here, equality is achieved iff $\delta(k)^2 +$ $\delta(l)^2$ is a constant for every $kl \in E(\Gamma)$.

For instance, the upper bound is achieved when Γ is regular, a star graph, etc.

Theorem 3.6. *Let* Γ *denote a graph with* $n \geq 3$. *Then*

$$NBSO(\Gamma) \leq \frac{mND_5(\Gamma) + {}^{nm}M_2F_N^*(\Gamma)}{2NSO(\Gamma)}, \tag{6}$$

where m is the size of Γ. *Here, equality is achieved iff* $\sqrt{\delta(k)^2 + \delta(l)^2}$ *is equal for all* $kl \in E(\Gamma)$.

Proof. Let $x_i = y_i = \sqrt{\delta(k)^2 + \delta(l)^2}$, $w_i = \frac{1}{\delta(k)\delta(l)}$ and $z_i = 1$, then using Lemma 2.3 we have,

$$m \sum_{kl \in E(\Gamma)} \frac{\delta(k)^2 + \delta(l)^2}{\delta(k)\delta(l)} + \sum_{kl \in E(\Gamma)} \frac{1}{\delta(k)\delta(l)} \sum_{kl \in E(\Gamma)} (\delta(k)^2 + \delta(l)^2)$$

$$\geq 2 \sum_{kl \in E(\Gamma)} \frac{\sqrt{\delta(k)^2 + \delta(l)^2}}{\delta(k)\delta(l)} \sum_{kl \in E(\Gamma)} \left(\sqrt{\delta(k)^2 + \delta(l)^2}\right),$$

$$\Rightarrow mND_5(\Gamma) + {}^{nm}M_2F_N^*(\Gamma) \geq 2NBSO(\Gamma)NSO(\Gamma).$$

which gives us the required inequality (6). Here, equality is achieved iff $\sqrt{\delta(k)^2 + \delta(l)^2}$ is equal for all $kl \in E(\Gamma)$.

3.2 Bounds of neighborhood modified Sombor index

Remark 1. The maximum value of the ratio $\frac{\delta(l)}{\delta(k)}$ in Γ, with $n \geq 3$ and for all pairs of vertices $k, l \in V(\Gamma)$ is achieved for maximum value of $\delta(l)$ and minimum value of $\delta(k)$. Therefore, for obtaining the maximum value of $\delta(l)$, we consider K_n. However, here $\frac{\delta(l)}{\delta(k)} = 1$. Thus, this case is disregarded. Now, let Γ be a link of K_n and P_2. A link of K_n and P_2 by the vertices $l \in V(K_n)$ and $s \in V(P_2)$, denoted by $K_n \sim P_2$ is defined by joining the nodes l and s with an edge in $K_n \in P_2$. The vertex $l \in V(K_n)$ has $\delta(l) = [(n-3)(\Delta-1)+2]$ in $K_n \sim P_2$, since all the vertices in its neighborhood have degree $n-1$ except for one vertex $s \in V(P_2)$ whose degree is 2 in $K_n \sim P_2$. The three vertices i.e. $l \in V(K_n)$, s, $k \in V(P_2)$ are the only vertices in $K_n \sim P_2$ with degree $\in \Delta - 1$. The vertex $k \in V(P_2)$ has the minimum nbd-degree sum such that $\delta(k) = 2$ in $K_n \sim P_2$.

Theorem 3.7. *If Γ denote a graph with $n \geq 3$. Then*

$$\frac{ND_4}{\sqrt{\frac{(n-3)(\Delta-1)+2}{2} + \frac{2}{(n-3)(\Delta-1)+2}}} \leq {}^mNSO(\Gamma) \leq \frac{1}{\sqrt{2}}ND_4. \tag{7}$$

Left equality is achieved if $G = K_n \sim P_2$, where Γ is a link of K_n and P_2 and right equality is achieved iff $\delta(k) = \delta(l)$ for all $kl \in E(\Gamma)$.

Proof. By arithmetic-geometric mean inequality,

$$\delta(k)^2 + \delta(l)^2 \geq 2\delta(k)\delta(l), \quad \sum_{kl\in E(\Gamma)} \frac{1}{\sqrt{\delta(k)^2 + \delta(l)^2}} \leq \sum_{kl\in E(\Gamma)} \frac{1}{\sqrt{2\delta(k)\delta(l)}},$$

which gives us the right inequality in (7). Here, equality is achieved iff $\delta(k) = \delta(l)$ for all $kl \in E(\Gamma)$. Now for the lower bound using Remark 1 we have,

$$\sum_{kl\in E(\Gamma)} \frac{1}{\sqrt{\delta(k)^2 + \delta(l)^2}} = \sum_{kl\in E(\Gamma)} \frac{1}{\sqrt{\delta(k)\delta(l)}\sqrt{\frac{\delta(k)}{\delta(l)} + \frac{\delta(l)}{\delta(k)}}}$$

$$\geq \sum_{kl\in E(\Gamma)} \frac{1}{\sqrt{\delta(k)\delta(l)}\sqrt{\frac{(n-3)(\Delta-1)+2}{2} + \frac{2}{(n-3)(\Delta-1)+2}}},$$

which gives us the left inequality in (7).

Theorem 3.8. *If Γ denote a graph with $n \geq 3$. Then,*

$$2\sqrt{2}\lambda \leq {}^mNSO(\Gamma) \leq \sqrt{m\lambda}. \tag{8}$$

where $\lambda = \sum_{kl\in E(\Gamma)} \frac{1}{\delta(k)^2+\delta(l)^2}$. Left equality is achieved if Γ is a disjoint union of P_3 and right equality is achieved iff $\delta(k)^2 + \delta(l)^2 = $ constant, for all $kl \in E(\Gamma)$.

Proof. Using Cauchy-Schwarz inequality we have,

$$^mNSO(\Gamma) = \sum_{kl \in E(\Gamma)} \sqrt{1}\sqrt{\frac{1}{\delta(k)^2 + \delta(l)^2}} \leq \sqrt{\sum_{kl \in E(\Gamma)} 1 \sum_{kl \in E(\Gamma)} \frac{1}{\delta(k)^2 + \delta(l)^2}},$$

which gives the right inequality (8). Here, equality is achieved iff $\delta(k)^2 + \delta(l)^2 =$ constant, for all $kl \in E(\Gamma)$. For instance, when Γ is a regular or a star graph. Now for the lower bound we have

$$^mNSO(\Gamma) = \sum_{kl \in E(\Gamma)} \sqrt{1}\sqrt{\frac{1}{\delta(k)^2 + \delta(l)^2}} = \sum_{kl \in E(\Gamma)} \frac{\sqrt{\delta(k)^2 + \delta(l)^2}}{\delta(k)^2 + \delta(l)^2}$$

$$\sum_{kl \in E(\Gamma)} \frac{\sqrt{2(2)^2}}{\delta(k)^2 + \delta(l)^2} = \sqrt{2}(2) \sum_{kl \in E(\Gamma)} \frac{1}{\delta(k)^2 + \delta(l)^2},$$

which gives us the left inequality in (8). Here, equality is achieved if Γ is a disjoint union of P_3, i.e., when $\delta(k) = \delta(l) = 2$ for all $kl \in E(\Gamma)$.

Theorem 3.9. *Let Γ denote a graph with $n \geq 3$. Then,*

$$\sqrt{2} \,^{mm}M_2(\Gamma) \leq \,^mNSO(\Gamma) \leq \frac{1}{\sqrt{2}}[(n-1)\Delta] \,^{nm}M_2(\Gamma). \tag{9}$$

Left equality is achieved if Γ is a disjoint union of P_3 and right equality is achieved if $G = K_n$.

Proof. We know,

$$^mNSO(\Gamma) = \sum_{kl \in E(\Gamma)} \frac{1}{\sqrt{\delta(k)^2 + \delta(l)^2}} = \sum_{kl \in E(\Gamma)} \frac{1}{\delta(k)\delta(l)} \frac{1}{\sqrt{\frac{1}{\delta(l)^2} + \frac{1}{\delta(k)^2}}}$$

$$\geq \sum_{kl \in E(\Gamma)} \frac{1}{\delta(k)\delta(l)} \frac{1}{\sqrt{\frac{1}{(2)^2} + \frac{1}{(2)^2}}} = \sum_{kl \in E(\Gamma)} \frac{1}{\delta(k)\delta(l)\sqrt{\frac{2}{(2)^2}}},$$

which gives us the left inequality in (9). Here, equality is achieved if Γ is a disjoint union of P_3, i.e., when $\delta(k) = \delta(l) = 2$ for all $kl \in E(\Gamma)$. Similarly, we can obtain the upper bound and hence the right equality is achieved for $\Gamma = K_n$, i.e., when $\delta(k) = \delta(l) = [(n-1)\Delta]$ for all $kl \in E(\Gamma)$.

Theorem 3.10. *If Γ denote a graph with $n \geq 3$, then*

$$^mNSO(\Gamma) + \,^mNSO(\bar{G}) \geq \frac{\sqrt{2}n}{4\Delta} \tag{10}$$

equality is achieved if $G = K_n$ or $G = \bar{K}_n$.

Proof. We know $|E(\Gamma)| + |\bar{E}(\bar{\Gamma})| = \,^nC_2$. Therefore,

$$^mNSO(\Gamma) + \,^mNSO(\bar{G}) = \sum_{kl \in E(\Gamma)} \frac{1}{\sqrt{\delta(k)^2 + \delta(l)^2}} + \sum_{kl \in E(\bar{G})} \frac{1}{\sqrt{\delta(k)^2 + \delta(l)^2}}$$

$$\geq \sum_{kl \in E(\Gamma)} \frac{1}{\sqrt{2}[(n-1)\Delta]} + \sum_{kl \in E(\bar{\Gamma})} \frac{1}{\sqrt{2}[(n-1)\Delta]} = \frac{^nC_2}{\sqrt{2}[(n-1)\Delta]},$$

which gives us the required inequality (10). Here, equality is achieved if $\Gamma = K_n$ or $\Gamma = K_n$.

Conclusion

A new neighborhood variants each of the Banhatti Sombor index and the modified Sombor index were defined in this study. Also, some bounds of these indices in terms of graph parameters like n, m, Δ, and other neighborhood degree-based topological indices like $^{nm}M_2$, $NH(\Gamma)$, ND_4, ND_5, F_N^*, NI and NSO were obtained. Moreover, characterization of the extremal graphs with respect to these indices was done in this article. The future scope of this work is that it can be extended to the computation of these indices for some important nanomaterials as well as to exploring mathematical relations between these indices and other existing topological indices.

Acknowledgments

The first author is very grateful to TMA Pai Foundation for extending research fellowship vide, letter No. 209/SMU/REG/TMAPURF/70/2020 dated 30th September 2021.

References

1. Adnan, Syed Ahtsham Ul Haq Bokhary, Abbas, G., and Iqbal, T. (2022). Degree-based topological indices and QSPR analysis of antituberculosis drugs. *J. Chem.*, 2022, 1–17.
2. Roćio, A.-S., Méndez-Bermúdez, J. A., Rodŕiguez, J. M., and Sigarreta, J. M.. (2021). Normalized Sombor indices as complexity measures of random networks. *Entropy*, 23(8), 976.
3. Roberto, C., Gutman, I., and Rada, J. (2021). Sombor index of chemical graphs. *Appl. Math. Comput.*, 399, 126018.
4. Das, K. C., Mondal, S., and Raza, Z. (2022). On Zagreb connection indices. *Eur. Phy. J. Plus*, 137(11), 1242.
5. de Julián-Ortiz, J. V. de, C de Gregorio Alapont, Rıos-Santamarina, I., Garcıa-Doménech, R., and Gálvez, J. (1998). Prediction of properties of chiral compounds by molecular topology. *J. Mol. Graph. Model.*, 16(1), 14–18.
6. Hanyuan, D., Tang, Z., and Wu, R. (2021). Molecular trees with extremal values of Sombor indices. *Int. J. Quant. Chem.*, 121(11), e26622.
7. Dragomir, S. S. (1984). On some inequalities (Romanian), Caiete Metodico Stiintific, Faculty of Mathematics, Timisoara University. *Romania*, 13.
8. Ivan, G. (2013). Degree-based topological indices. *Croatica Chemica Acta.*, 86(4), 351–361.
9. Ivan, G. (2021). Geometric approach to degree-based topological indices: Sombor in- dices. *MATCH Commun. Math. Comput. Chem.*, 86(1), 11–16.
10. Frank, H. (1969). *Graph Theory*. Reading, MA: Addison-Wesley.
11. Yufei, H. and Liu, H. (2021). Bounds of modified Sombor index, spectral radius and energy. *AIMS Math.*, 6(10), 11263–11274.
12. Kulli, V. R. (2021a). Neighborhood Sombor index of some nanostructures. *Int. J. Math. Trends Technol.*, 67(5), 101–108.

13. Kulli, V. R. (2021b). On Banhatti-Sombor indices. *SSRG Int. J. Appl. Chem.*, 8(1), 21–25.
14. Kwun, Y. C., Munir, M., Nazeer, W., Rafique, S., and Kang, S. M. (2017). M-polynomials and topological indices of V-phenylenic nanotubes and nanotori. *Scientific Reports* 7(1), 1–9.
15. Zhen, L., Zhou, T., Kulli, V. R., and Miao, L. (2021). On the first Banhatti- Sombor index. *J. Int. Math. Virtual Inst.*, 11(1), 53–68.
16. Sourav, M., Dey, A., De, N., and Pal, A. (2021). QSPR analysis of some novel neighbourhood degree-based topological descriptors. *Complex Intel. Sys.* 7(2), 977–996.
17. Sourav, M., Some, B., Pal, A., and Das, K. C. (2022). On neighborhood inverse sum indeg energy of molecular graphs. *Symmetry*, 14(10), 2147.
18. Nadeem, M. F., Zafar, S., and Zahid, Z. (2016). On topological properties of the line graphs of subdivision graphs of certain nanostructures. *Appl. Math. Comput.*, 273, 125–130.
19. Izudin, R. (2021). Chemical applicability of Sombor indices. J. Serbian Chem. Soc., 86(5), 445–457.
20. Izudin, R. and Furtula, B. (2020). Predictive potential of eigenvalue-based topological molecular descriptors. *Journal of Computer-Aided Molecular Design.*, 34(9), 975–982.
21. Tamás, R., Dŏslic, T., and Ali, A. (2021). On the Sombor index of graphs. *Contrib. Math.*, 3, 11–18.
22. Weiner, H. (1947). Prediction of isomeric difference in para□n properties. *J. Am. Chem. Soc.*, 69, 17–20.
23. Xuewu, Z., Rather, B. A., Imran, M., and Ali, A. (2022). On some topological indices defined via the modified Sombor matrix. *Molecules*, 27(19), 6772.

16 Performance of transverse surface roughness on conical bearing lubricated with ferrofluid in the existence of couple stresses

Yogini D. Vashi[1,a], Rakesh M. Patel[2,b] and Gunamani B. Deheri[3,c]

[1]Assistant Professor, School of Computer Science, Artificial Intelligence & Emerging Technologies, TransStadia University, Ahmedabad-380022, Gujarat, India

[2]Associate Professor, Department of Mathematics, Gujarat Arts and Science College, Ahmedabad-380006, Gujarat, India

[3]Former Associate Professor, Department of Mathematics, Sardar Patel University, Vallabh Vidyanagar-388120, Gujarat, India

Abstract

This manuscript presents a theoretical examination of how transverse surface irregularities impact conical bearing using ferrofluid as a lubricant, considering the presence of couple stresses. The Christensen and Tonder stochastic averaging model is employed to describe the statistical characteristics of transverse surface roughness, encompassing non-zero mean, variance and skewness. The magnetic fluid flow effects are captured through the well-established Neuringer-Roseinweig model, while the impact of couple stresses is analyzed using the Stoke's microcontinuum theory. The equation of the pressure distribution is achieved from the stochastically averaged modified Reynolds equation, enabling the computation of load-bearing capacity. Graphical representations are provided for various dimensionless parameters. The results highlight the pronounced combined impact of magnetization and couple stresses, particularly when negatively skewed roughness is present and the cone's half vertical angle is at its lowest, as compared to the nonmagnetic situation.

Keywords: Ferrofluid, load bearing capacity, couple stresses, rough conical bearing

Introduction

The phenomena of squeeze film lubrication have been expansively demonstrated in various manufacturing and engineering applications including engine parts, gears, bearings, lubricating couplings and more. Fundamentals of fluid film lubrication are studied by Cameron (1981), Hamrock (1994) with Newtonian fluids. Nowadays lubricant with additives such as microploar fluid, couple stress fluid, and ferrofluid is becoming of huge interest owing to its non-Newtonian performance. Ferrofluid plays a very crucial role in squeeze film lubrication since the magnetic fluid has important applications in manufacturing. These fluids have been accountable for the improvement and development of several new appliances and devices in the areas of energy renovation, computers, hi-fi speakers,

[a]yogini.vashi@gmail.com, [b]rmpatel12711@gmail.com, [c]gm.deheri@rediffmail.com

semiconductors and space research. Their key property is that they can be prepared to follow to any place desired through the aid of magnets and to move even
in zero gravity states, and hence due to this reason, much research work is carried
out on magnetic fluid (Verma, 1986; Kumar et al., 1992; Mehta and Upadhyay,
1999; Bhat, 2003; Patel et al., 2007; Shah et al., 2016). Three essential model
based on magnetic fluid are extensively studied by Huang and Wang (2016). In
unconventional scenarios, it is expected that bearing surfaces should possess a
smooth texture. However, these bearing surfaces develop some roughness over
time due to factors like running, wear, and tear. Recently, there has been a significant focus on examining how surface roughness impacts the performance of
bearings. Several surface models have been utilized to investigate the importance
of surface roughness, taking into account the presence of squeeze film lubrication.
The first model was developed by Christensen and Tonder (1969a, b, 1970). They
have considered patterns of surface roughness both, longitudinally and transversely by implementing the function of probability density.

Many researchers have employed the stochastic approach given by Christensen
and Tonder (1969a, b, 1970) for different bearing geometries. For example, Deheri
et al. (2013) studied transverse roughness's effectiveness in the context of conical
bearing. The hydromagnetic rough conical bearing was deliberated by Vadhere et
al. (2010). The enactment of longitudinally rough conical bearing was examined
by Andharia and Deheri (2010). The enactment of longitudinally rough conical
bearing with considering slip boundary conditions studied by Patel and Deheri
(2016). All of these studies revealed that in the presence of ferrofluid, along with
surface roughness that is skewed to the negative having the minimum angle for
the semi vertical of the cone, there is significant improvement in the load bearing capacity when compared with the case of conventional lubricant. Lin et al.
(2012) examined the performance of conical plates with squeeze film in between
when used with micropolar fluid and revealed that load bearing capacity was
boosted due to micropolar fluid. Nowadays many researchers are using a couple
stress types of fluids as lubricants because of its non-Newtonian behavior. In literature, Ramanaiah (1979) studied finite plates that had squeeze film between
them which used couple stress fluid for lubrication. Their findings showed an
increase in load bearing capacity when couple stress fluid is used. Naduvinamani
et al. (2016) presented a study about conical plates with squeeze film in between
using viscosity variation when used with couple stresses. Their findings suggested
an improvement in squeeze film time as well as load bearing capacity because of
pressure dependent viscosity and couple stress fluid. Hanumagowda et al. (2021)
studied the added impact of surface roughness and magnetic field on porous conical bearing which used couple stress fluid as lubricant. Their findings suggested
an increase in squeeze film time and the capacity of load bearing due to couple
stresses, magnetic field and surface roughness as opposed to the Newtonian case
and the non-magnetic case. Rao and Rahul (2019) presented the Rabinowitsch
fluid model for rough conical bearing with variable viscosity. They discovered
that pressure and load bearing capacity increased due to surface roughness while
a variation of viscosity affects adversely on pressure distribution and hence load
bearing capacity. In 2018, Daliri and Javani studied conical plates, specifically
the squeezing motion amid them when ferrofluid couple stress types of lubricants

are used with the impact of convective fluid inertia. They have observed a positive influence on bearing performance of ferrofluid couple stress lubricants as well as convective fluid inertia parameters. Vashi et al. (2018) depicted the study of porous and rough stepped plates which are lubricated using ferrofluid along with couple stresses. Additionally, Vashi et al., (2020) studied porous and rough circular stepped plates when used with ferrofluid couple stress type of lubricant. Both these studies showed that the added impact of couple stresses and ferrofluid when used with surface roughness that is negatively skewed made the load bearing capacity better in comparison to non-magnetic case.

According to the past scrutiny reviewed above, the enactment of rough conical plates using ferrofluid as a lubricant, considered along with couple stresses is not studied accurately yet. The present analysis attempts to study the joint impact that couple stresses and ferrofluid have on rough conical bearing by using Neuringer and Roseinweig's ferrofluid model (1964) and Stokes' couple stress fluid model (1996).

Analysis

In Figure 16.1, the configuration of rough conical plates in a squeeze film setup is depicted, where couple stresses are taken into account while utilizing ferrofluid as a lubricant. The geometrical arrangement is subjected to an oblique magnetic field. The thickness of the fluid film along the cone axis is denoted as h, while the cone's half vertical angle is represented by ω and the conical plate's radius is indicated by the a. The bearing surface is assumed to be rough in a transverse manner. All the foundational assumptions of hydrodynamic lubrication, as applied in the theory of thin film lubrication, are upheld in this current investigation. The governing equations for the motion, following the principles of hydrodynamic

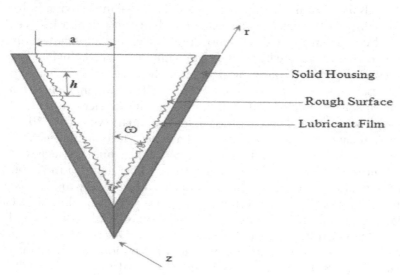

Figure 16.1 Physical structure of bearing system

lubrication, have been developed for the fluid flow with couple stresses in the film region by Stokes (1966).

$$\mu\frac{\partial^2 u}{\partial z^2} - \eta\frac{\partial^4 u}{\partial z^4} = \frac{\partial p}{\partial r} \tag{1}$$

$$\frac{\partial p}{\partial z} = 0 \tag{2}$$

$$\frac{1}{r}\frac{\partial}{\partial r}(ru) + \frac{\partial w}{\partial z} = 0 \tag{3}$$

In the above equations, the velocity components in the r and z directions are symbolized by u and w correspondingly. The fluid film pressure is denoted as p, while the dynamic viscosity of the lubricant is characterized by μ. Additionally the constant η is attributed to the material properties accountable for the behavior of couple stress fluids. Regarding the velocity components, the corresponding boundary conditions are defined as follows:

$$u=0 \text{ and } \frac{\partial^2 u}{\partial y^2} = 0 \text{ with } z=0 \text{ and } z=h\sin\omega \tag{4}$$

$$w=0 \text{ at } z=0 \tag{5a}$$

$$w=h^\bullet\sin\omega \text{ at } z=h\sin\omega \tag{5b}$$

The solution of the equation (1) is obtained through the help of the boundary condition given in the equation (4) as below

$$u(r,z) = \frac{1}{2\mu}\frac{\partial p}{\partial r}\left[z^2 - zh\sin\omega + 2l^2 - 2l^2\frac{\cosh(2z - h\sin\omega)}{\cosh\left(\dfrac{h\sin\omega}{2l}\right)}\right] \tag{6}$$

Utilizing the equation (6) within equation (3) and integrating over the fluid film thickness, while in view of the provided boundary conditions as outlined in equation (5a) and equation (5b) offers the modified Reynolds-type equation for smooth conical bearing with couple stress fluid

$$\frac{1}{r}\frac{d}{dr}\left[f_1(h,l)r\frac{dp}{dr}\right] = 12\mu h^\bullet\sin\omega \tag{7}$$

Where $f_1(h,l) = h^3\sin^3\omega - 12l^2 h + 24l^3\tanh\left(\dfrac{h\sin\omega}{2l}\right)$. \hfill (8)

In accordance with Neuringer and Rosensweig (1964) model equation (7) turns out to be as below

$$\frac{1}{r}\frac{d}{dr}\left[f_1(h,l)r\frac{d}{dr}\left(p-\frac{\mu_0\bar{\mu}H^2}{2}\right)\right]=12\mu\,\dot{h}^\bullet\sin\omega.$$

(9)

Where the magnetic field's magnitude H is defined as

$$H^2=A\left(r^2-a^2\cos ec^2\omega\right).$$

(10)

The fluid pressure's boundary conditions are

$$p(a\cos ec\omega)=0\,,\left(\frac{dp}{dr}\right)_{r=0}=0.$$

(11)

Developing the averaging practice of Christensen and Tonder (1969 a, b, 1970) deliberated by Deheri et al. (2013). Equation (8) becomes:

$$\frac{1}{r}\frac{d}{dr}\left[f_1(h,\sigma,\alpha,\ \varepsilon,l)r\frac{d}{dr}\left(p-\frac{\overline{\mu_0\bar{\mu}H^2}}{2}\right)\right]=12\mu\dot{h}^\bullet\sin\omega.$$

(12)

Where,

$$f_1(h,\sigma,\alpha,\varepsilon,l)=h^3\sin^3\omega+3\sigma^2h\sin\omega+3h^2\alpha\sin^2\omega+3\alpha^2h\sin\omega+3\sigma^2\alpha+\alpha^3+\varepsilon$$
$$-12l^2h\sin\omega+24l^3\tanh\left(\frac{h\sin\omega}{2l}\right)$$

By taking integration of equation (9) with respect to r and operating the pressure boundary conditions provided by equation (11), we arrive at the expression for the fluid film pressure as presented below.

$$p=\frac{3\mu\sin\omega\dot{h}^\bullet\left(r^2-a^2\cos ec^2\omega\right)}{f_1(h,\sigma,\alpha,\varepsilon,l)}+\frac{\mu_0\bar{\mu}H^2}{2}$$

(13)

Expression for dimensionless pressure is obtained from the below equation

$$P=-\frac{h^3p}{\mu\dot{h}^\bullet\pi a^2\cos ec\omega}$$

(14)

$$P=\frac{3}{\pi F(\sigma^*,\alpha^*,\varepsilon^*,l^*)}\left(1-R^2\right)+\frac{\mu^*\cos ec\omega}{2\pi}\left(1-R^2\right)$$

(15)

Where

$$F\left(\sigma^*,\alpha^*,\varepsilon^*,\iota^*\right)=\sin^3\omega+3\sigma^{*2}\sin\omega+3\alpha^*\sin^2\omega+3\alpha^{*2}\sin\omega+3\sigma^{*2}\alpha^*$$
$$+\alpha^{*3}+\varepsilon^*-3\iota^*\sin\omega+3\iota^{*3}\tanh\left(\frac{\sin\omega}{\iota^*}\right)$$

(16)

Non-dimensional parameters present in equation (15) defined as below:

$$\sigma^*=\frac{\sigma}{h}\ ,\ \alpha^*=\frac{\alpha}{h}\ ,\ \varepsilon^*=\frac{\varepsilon}{h^3}\ ,\ \iota^*=\frac{2\iota}{h}\ ,\ \mu^*=-\frac{\mu_0\,\overline{\mu}h^3A}{\mu h^\bullet}\ ,\ R=\frac{r}{a\cos ec\omega}$$

Equation of the bearing's load capacity is found by integrating the film pressure across the film region as follows:

$$W=\int_0^{a\cos ec\omega}2\pi r\,p(r)dr$$

(17)

Dimensionless load bearing capacity can be obtained from the below equation

$$W^*=-\frac{h^3W}{\mu h^\bullet\pi a^4\cos ec^2\omega}$$

(18)

$$W^*=\frac{3}{2}\frac{\cos ec\omega}{F\left(\sigma^*,\alpha^*,\varepsilon^*,\iota^*\right)}+\frac{\mu^*}{4}\cos ec^2\omega$$

(19)

Results and discussion

The present article represents transverse surface roughness's performance on conical bearing lubricated with ferrofluid in the existence of couple stresses. The impact caused by magnetic fluid can be perceived using the dimensionless magnetization parameter μ^*. Transverse surface roughness is described by the dimensionless parameters α^* (non-zero mean), ε^* (variance) and σ^* (standard deviation). The dimensionless parameter ι^* offers the influence of couple stresses. From equation (19), it can be perceived that $\frac{\mu^*}{4}\cos ec^2\omega$ times enhancement is found in load bearing capacity as compared to the bearing system based on Newtonian lubricant. The trends of dimensionless load capacity of bearing with magnetization parameter μ^* for distinct values of σ^*, α^*, ε^*, ι^* and ω is displayed from Figures 16.2–16.6. It suggests a growth in load carrying capacity along with a rise of the magnetization parameter's value. It is noted that parallel to an increase in the value of magnetization parameter, the load carrying capacity also increases. It displays that maximum load will be registered in couple stress parameter's case, which happens because of the polar additives

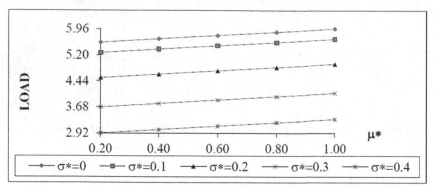

Figure 16.2 Change in load with regards to μ^* considering distinct values of σ^* along with $\alpha^* = 0.05$, $\varepsilon^* = 0.05$, $\iota^* = 0.20$, and $\omega = 45$

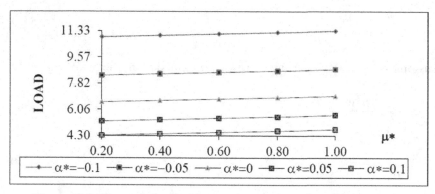

Figure 16.3 Change in load with regards to μ^* considering distinct values of α^* along with $\sigma^* = 0.10$, $\varepsilon^* = 0.05$, $\iota^* = 0.20$, and $\omega = 45$

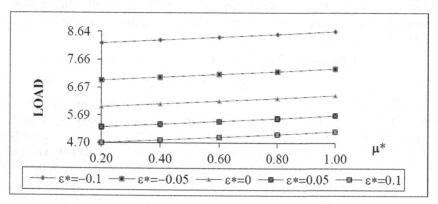

Figure 16.4 Change in load with regards to μ^* considering distinct values of ε^* along with $\sigma^* = 0.10$, $\alpha^* = 0.05$, $\iota^* = 0.20$, and $\omega = 45$

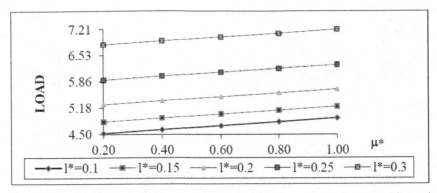

Figure 16.5 Change in load with regards to μ^* considering distinct values of ι^* along with $\sigma^* = 0.10$, $\alpha^* = 0.05$, $\varepsilon^* = 0.20$, and $\omega = 45$

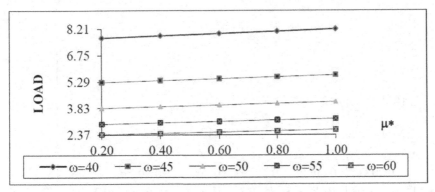

Figure 16.6 Change in load with regards to μ^* considering distinct values of ω along with $\sigma^* = 0.10$, $\alpha^* = 0.05$, $\iota^* = 0.20$, and $\varepsilon^* = 0.05$

chain length in the case of a non-polar lubricant. Therefore, the parameter of couple stress displays the workings of bearing geometry lubricant's interface.

Figures 16.7–16.9 demonstrate load distribution in the context of σ^* for different values of α^*, ι^* and ω. All these figures display the negative impact of σ^*. The bearing surface's roughness is in conflict with the fluid flow motion of bearing which results into reduction in fluid film pressure and hence the reduction will take place in the bearing load capacity. Figures 16.7 and 16.8 confirmed that the contrary influence of σ^* may reduce till some extent with the joint effect of ι^* and μ^*.

Figures 16.10 and 16.11 exhibit load distribution with reference to α^* for varying values of ι^* and ω. Both these figures witnessed that positive increase in α^* will lead to a depreciation of load capacity while opposite trend is suggested with surface roughness skewed negatively. From Figure 16.10, it is seen that maximum load is noted for large values of ι^* with suitable choice of μ^*. Figures 16.12–16.15 represents load distribution with reference to ι^* for varying values of μ^*, α^*, σ^* and ω. All these figures display the positive impact of the parameter

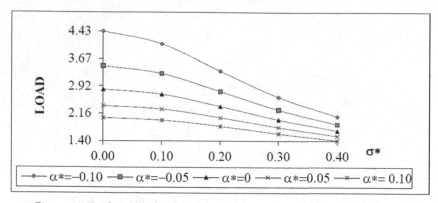

Figure 16.7 Change in load with regards to σ^* considering distinct values of α^* along with $\varepsilon^* = 0.05$, $\mu^* = 0.40$, $\iota^* = 0.20$, and $\omega = 45$

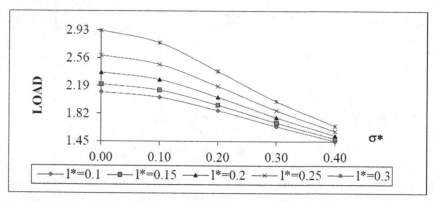

Figure 16.8 Change in load with regards to σ^* considering distinct values of ι^* along with $\varepsilon^* = 0.05$, $\mu^* = 0.40$, $\alpha^* = 0.05$, and $\omega = 45$

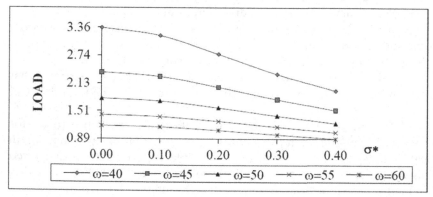

Figure 16.9 Change in load with regards to σ^* considering distinct values of ω along with $\varepsilon^* = 0.05$, $\mu^* = 0.40$, $\alpha^* = 0.05$, and $\iota^* = 0.20$

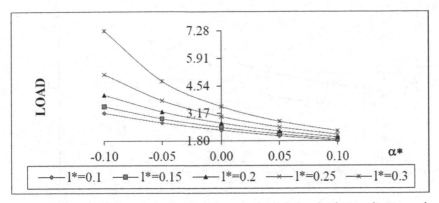

Figure 16.10 Change in load with regards to α^* considering distinct values of ι^* along with $\varepsilon^* = 0.05$, $\mu^* = 0.40$, $\sigma^* = 0.10$, and $\omega = 45$

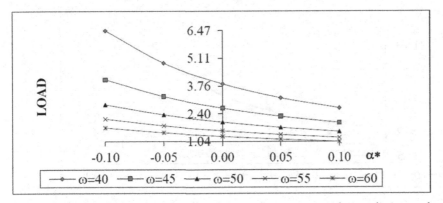

Figure 16.11 Change in load with regards to α^* considering distinct values of ω along with $\varepsilon^* = 0.05$, $\mu^* = 0.40$, $\sigma^* = 0.10$, and $\iota^* = 0.20$

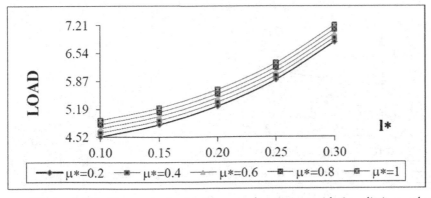

Figure 16.12 Change in load with regards to ι^* considering distinct values of μ^* along with $\varepsilon^* = 0.05$, $\alpha^* = 0.05$, $\sigma^* = 0.10$, and $\omega = 45$

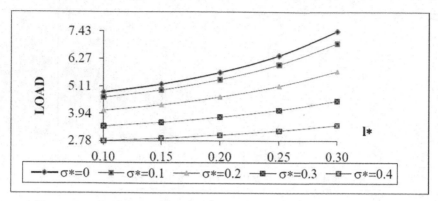

Figure 16.13 Change in load with regards to ι^* considering distinct values of σ^* along with $\varepsilon^* = 0.05$, $\alpha^* = 0.05$, $\mu^* = 0.40$, and $\omega = 45$

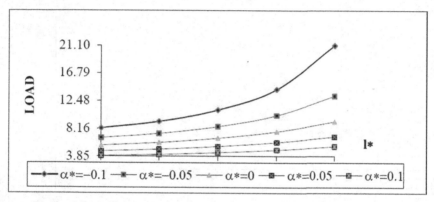

Figure 16.14 Change in load with regards to ι^* considering distinct values of α^* along with $\varepsilon^* = 0.05$, $\sigma^* = 0.10$, $\mu^* = 0.40$, and $\omega = 45$

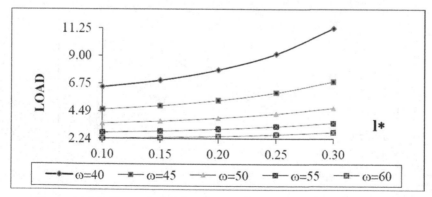

Figure 16.15 Change in load with regards to ι^* considering distinct values of ω along with $\varepsilon^* = 0.05$, $\sigma^* = 0.10$, $\mu^* = 0.40$, and $\alpha^* = 0.05$

of couple stress as the value of ι^* increases parallelly to an increase in the load. It is perceived that maximum load is achieved in negatively skewed roughness's case for large value of ι^*.

Conclusion

This article posits a theoretical examination of performance of transverse surface roughness on conical bearing lubricated with ferrofluid in the existence of couple stresses. Combined impact of ferrofluid and couple stresses plays a vital role in bearing's load capacity. Based on the graphical presentation we have attained the following conclusions.

1. When couple stresses are introduced, the utilization of ferrofluid lubricant leads to an enhanced load capacity as opposed to the case of conventional lubricant.
2. Through the optimal interplay of couple stress and magnetization parameter, having surface roughness that is skewed negatively the performance of rough conical plates enhances, especially when the cone's half vertical angle is minimized.
3. The decrease in bearing's load capacity because of standard deviation and positively skewed roughness can be mitigated to a certain extent by employing ferrofluid lubricant in conjunction with the couple stress parameter. This approach can effectively extend the lifespan of the rough conical bearing.

References

1. Cameron, A. (1981). Basic lubrication theory. New York: Wiley.
2. Hamrock, B. J. (1994). Fundamentals of fluid film lubrication. New York: McGraw-Hill.
3. Verma, P. D. S. (1986). Magnetic fluid-based squeeze film. *Int. J. Eng. Sci.*, 24, 395–401.
4. Shah, R. C., Patel, N. I., and Kataria, R. C. (2016). Some porous squeeze film-bearings using ferrofluid lubricant a review with contributions. *Arc. Proc. Ins. Mech. Eng. Part J J. Engg. Tribol.*, 230(9), 1157–1171. DOI: 10.1177/1350650116629096.
5. Mehta, R. V. and Upadhyay, R. V. (1999). Science and technology of ferrofluids. *Curr. Sci.*, 76(3), 305–312.
6. Bhat, M. V. (2003). Lubrication with a magnetic fluid. Team Spirit (India) Pvt. Ltd.
7. Kumar, D., Sinha, P., and Chandra, P. (1992). Ferrofluid Squeeze film for spherical and conical bearings. *Int. J. Engg. Sci.*, 30(5), 645–656.
8. Patel, R. M. and Deheri, G. M. (2007). Magnetic fluid based squeeze film between porous conical plates. *Indus. Lub. Tribol.*, 59(3), 309–322.
9. Huang, W. and Wang, X. (2016). Ferrofluids lubrication a status report. *Lub. Sci.*, 28, 3–26.
10. Christensen, H. and Tonder, K. C. (1969a). Tribology of rough surface: Stochastic models of hydrodynamic Lubrication. *SINTEF*, Report No: 10/69-18.
11. Christensen. H. and Tonder. K. C. (1969b). Tribology of rough surfaces: Parametric study and comparison of lubrication models. *SINTEF*, Report No: 22/69-18.

12. Christensen, H. and Tonder, K. C. (1970). The hydrodynamic lubrication of rough bearing surfaces of finite width. *ASME-ASLE Lub. Conf.*, Cincinnati, Ohio, Paper no. 70-Lub-7.

13. Deheri, G. M., Patel, R. M., and Patel, H. C. (2013). Magnetic fluid based squeeze film between porous rough conical plates. *J. Comput. Methods Sci. Engg.*, 13, 419–432.

14. Vadher, P. A., Deheri, G. M., and Patel, R. M. (2010). Performance of a hydromagnetic squeeze films between conducting porous rough conical plates. *Mechanica*, 45, 767–783. DOI: 10.1007/s11012-010-9279-y.

15. Andharia, P. I. and Deheri, G. M. (2010). Longitudinal roughness effect on magnetic fluid-based squeeze film between conical plates. *Indus. Lub. Tribol.*, 62(5), 285–291. DOI: 10.1108/00368791011064446.

16. Patel, J. R. and Deheri, G. M. (2016). The effect of slip velocity on the ferrofluid based film in longitudinally rough conical plates. *J. Serbian Soc. Comput. Mech.*, 10(2), 18–29.

17. Neuringer, J. L. and Rosensweig, R. E. (1964). Magnetic fluids. *Phy. Fluids.*, 7(12), 1927.

18. Lin, J. R., Kuo, C. C., Liao, W. H., and Yang, C. B. (2012). Non Newtonian micropolar fluid squeeze film between conical plates. *Z. Naturforsch*, 67(a), 333–337.

19. Ramanaiah, G. (1979). Squeeze film between finite plates lubricated by fluids with couple stress. *Wear*, 54, 315–320. DOI: 10.1016/0043-1648(79)90123-6.

20. Naduvinamani, N. B., Siddangouda, A., and Ayyappa, G. H. (2016). Squeeze film characteristics of conical bearings with combined effects of piezo-viscous dependency and non-Newtonian couple stresses. *Tribol. Mat. Surf. Interf.*, 10(3), 126–130, DOI: 10.1080/17515831.2016.1235842.

21. Hanumagowda, B. N., Salma, A., and Nair, S. S. (2021). Combined effect of rough surface with MHD on porous conical bearing with conducting couple-stress fluid. *Palestine J. Math.*, 10, 59–68.

22. Rao, P. S. and Rahul, A. K. (2019). Pressure generation in rough conical bearing using non-Newtonian Rabinowitsch fluid with variable viscosity. *Indus. Lub. Tribol.*, 71(3), 357–365. DOI: 10.1108/ILT-01-2018-0035.

23. Stokes, V. K. (1966). Couple-stresses in fluids. *Phys. Fluids.*, 9(9), 1709–1715

24. Daliri, M. and Javani, N. (2018). Squeeze film characteristics of ferro-coupled stress inertial fluids in conical plates. *Indus. Lub. Tribol.*, 70(5), 872–877. DOI: 10.1108/ILT-02-2017-0047.

25. Vashi, Y., Patel, R. M., and Deheri, G. M. (2018). Ferrofluid based squeeze film lubrication between rough stepped plates with couple stress effect. *J. Appl. Fluid Mech.*, 11(3), 597–612, DOI:10.29252/jafm.11.03.27854.

26. Vashi, Y., Patel, R., and Deheri, G. M. (2020). Neuringer-roseinweig model based longitudinally rough porous circular stepped plates in the existence of couple stress. *Acta Polytech.*, 60(3), 259–267. DOI:10.14311/AP.2020.60.0259.

17 Analysis of psychological distress among the college students using the concept of fuzzy logic system

Monika Rathore[1,a], Uday Raj Singh[1] and Sanjeev Kumar[2]

[1]Department of Mathematics, C. L. Jain College, Firozabad, India

[2]Department of Mathematics, Dr. Bhimrao Ambedkar University, Khandari Campus, Agra, India

Abstract

COVID-19 had caused so many disturbances to continue a normal life in many aspects. One of them is that it had led to an unavoidable surge in the use of modern technology among college students to continue their studies and other daily works. Absolutely, electronic gadgets for e.g. cell phones, laptops, and personal computers are now become an integral and important part of students' accessories in this modern era which endangers psycho-social health of students. As an outcome, radiations from these devices caused psycho-social complications such as – aggression, anxiety, depression, self-harm, eating disorders, isolation, etc., which finally develop a high risk of negative impact on students' mental health. Mental health among college students has become significant and growing concern that appears to be increasing over time. Therefore, the main intention of this proposed work is to analyze the psychological distress by applying the concept of fuzzy logic system, which caused due to the extensive use of modern technology.

Keywords: Fuzzy logic, modern technology, psychological distress, psycho-social problem, students' well-being

Introduction

The COVID-19 pandemic had made many disturbances and complications in many phases to live a normal life. The extended lockdown and online classes have increased the screening time of college students. In recent years, the worldwide issue that has been amplified during and after Covid-19 is the excessive use of electronic gadgets or modern technology among college students to attain different learning skills, to join online classes, and for other daily works. Unfortunately, this excessive use of modern technology reveals the detrimental impact on the life of college students which harms psycho-social health and develops the risk of various mental and physical illnesses like – aggression, anxiety, depression, distraction, dehumanization, headache, social isolation, isolated feelings, shorter attention span, sleep disorder, eating disorder, self-harm ideation, vision problem, and anti-social behavior, etc. Many studies unveiled that in recent years – aggression, anxiety, depression, anti-social behavior, self-harm ideation, and eating disorder are reported as the most common psycho-social factors which drastically

---

[a]monikarathore16sep@gmail.com

affected the psycho-social health of college students and as a result, their academic performance and health is being disturbed which caused psychological distress among them.

Literature review

Several studies have shown that in today's world, technology plays a significant role which is increasing day by day, and the pandemic highlighted how students are dependent on technology. Studying and working with electronic gadgets has become the order of the day and as a result, college students suffered and suffering from psychological distress, which is a major concern related to the health of college students. In this direction to analyze the level of psychological distress among college students, Diomidous and Mantas (2016) showed that information technology offers quick access and provides a facility to communicate anywhere but is dangerous for health. So, users should be aware of its excessive use. Hoge et al. (2017) confirmed the relationship between digital media and depression. They showed the use of electronic media is associated with depressive symptoms and indicated that in some situations, the utilization of the nature of digital talk can be helpful to improve mood and upgrade health-enhancing strategies. Joshi (2021) established a reciprocal correlation between psychosocial distress and development. The author highlighted the need to purposefully integrate mental health concerns into the development response. In a diathesis-stress framework, Morrison and Connor (2005) investigated the interaction between rumination and different stress measures which could predict psychological distress components. Nguyen et al. (2020) presented the work which showed the connection between social sites addiction and mental health disorders. Nsereko et al. (2014) aimed to address the multi-dimensional aspect of psychosocial complications in the development of psychopathology for the students of Ugandan University. They hypothesized that there is no relationship between psychosocial problems and the enlargement of psychopathology. Pachiyappan et al. (2021) aimed to determine the dissimilarity in duration of using electronic devices before and during the pandemic. They also find out the associated health impacts.

According to World Health Organization (WHO), the most important basis of a student's life is good mental health and well-being. So, it is one of the significant psycho-social health concerns to be analyzed for the well-being of college students. In the proposed work five major factors are taken as input, which are contributing to psychological distress. Psychological distress is widely used for indicating mental health but remains vague to understand. This work presents a methodology to analyze psychological distress by utilizing the concept of fuzzy logic system and showed that the use of fuzzy logic system provides an optimum platform to remove vagueness and deal with psychological distress among college students caused due to excessive use of modern technology.

Psychological distress

Psychological distress encloses the signs and experiences of an individual's internal life that are disturbing, confusing, complicated, and strange. It gives rise to unusual behavior, and infects a person's sensations, performance, and relations with people around them. Therefore, psychological distress covers a wide range from normal feelings like unhappiness and fear to certain traumatic life experiences such as aggression, anxiety, depression, worries, hopelessness, eating disorders, addictive behavior, social isolation, etc.

Methods to diagnose psychological distress
The different diagnostic methods for psychological distress are following:

a)　**A physical checkup:** In this method, healthcare tries to rule out physical problems that cause the symptoms of psychological distress.
b)　**Laboratory tests:** This method includes blood tests, that indicate whether a physiological state such as thyroid or an electrolyte disproportion is causing psychological distress, or screening for alcohol and drugs may also done.
c)　**Psychological assessments:** In this method, the healthcare professional asks about the symptoms, thoughts, feelings, and behavior or asks to fill out a questionnaire to get answers to these questions.

Methodology

In this age of information technology, the debilitating effects on mental and physical health due to excessive use of modern technology are proven as a growing worldwide concern. The purposed work is aimed to analyze the level of psychological distress among college students by using the fuzzy logic system. In this process for data collection, a questionnaire was designed and surveyed in offline mode among 300 students of some colleges affiliated to Dr. Bhimrao Ambedkar University Agra. In the process of this work, the basic structure of the proposed work is shown in Figure 17.1.

There are so many factors causing psychological distress but on the basis of investigation, the most prominent factors affecting psycho-social health are-aggression, anxiety, depression, social behavior, and self-harm ideation are taken as input and psychological distress as output in the fuzzy logic system. "Fuzzy logic is an extension of two-valued logic, it facilitates dealing with imprecise reasoning, and vague propositions and it is widely used as a basis for decision

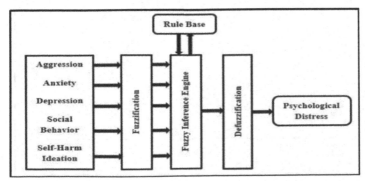

Figure 17.1 Basic structure of proposed model using fuzzy logic system

analysis to handle the concept of truth, where the values of truth range between completely true and completely false" Zadeh, L. A. (1965).

All inputs factors are very significant in the proposed model, now for fuzzification, input factors are classified into three fuzzy sets from which three inputs have – Low, Medium, High and fourth input has – Excellent, Good, and Poor, and fifth input has – Rare, Often, and Always fuzzy sets. While, output is divided into five fuzzy sets – Very Low, Low, Moderate, High, Very High by using the triangular membership function. The triangular membership function in fuzzy logic is as follow given by equation (1).

$$\mu_{mf}(z) = \begin{cases} 0 & z < \alpha^* \\ \frac{z-\alpha^*}{\beta^*-\alpha^*} & \alpha^* \le z \le \beta^* \\ \frac{\gamma^*-z}{\gamma^*-\beta^*} & \beta^* \le z \le \gamma^* \\ 0 & z > \gamma^* \end{cases} \tag{1}$$

After fuzzification, fuzzy logic is implemented to analyze psychological distress. Figure 17.2 shows the proposed fuzzy inference system.

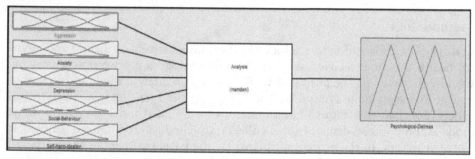

Figure 17.2 Proposed FIS system for analysis

Figures 17.3–17.8 present triangular membership function plotting of input and output variables, which are shown below.

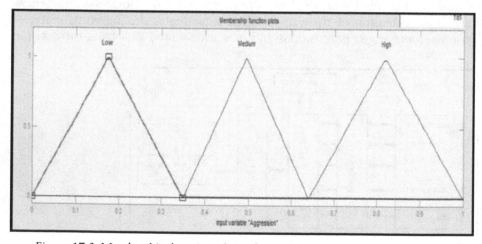

Figure 17.3 Membership function plots of aggression

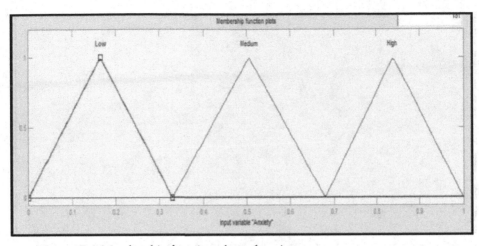

Figure 17.4 Membership function plots of anxiety

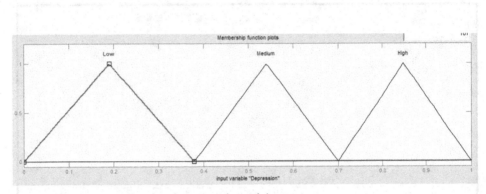

Figure 17.5 Membership function plots of depression

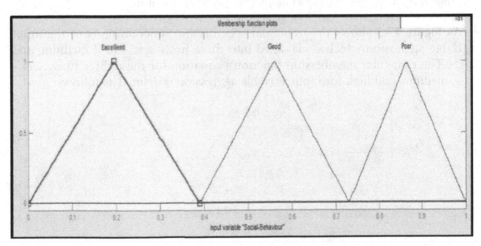

Figure 17.6 Membership function plots of social-behavior

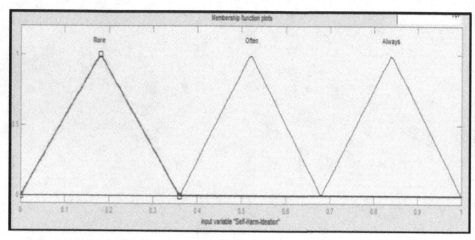

Figure 17.7 Membership function plots of self-harm ideation

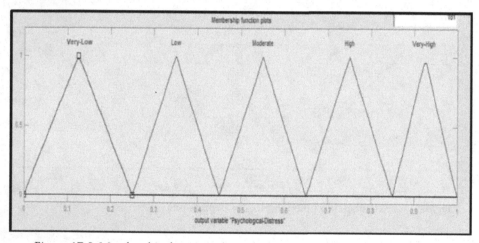

Figure 17.8 Membership function plots of psychological distress

As Figure 17.3 shows the triangular membership function plots of the input variable aggression which is classified into three fuzzy sets – low, medium, and high. The triangular membership function's equation for these three fuzzy sets – low, medium, and high for input variable aggression is defined as follows:

$$\mu_{Low}(z_1) = \begin{cases} 0 & z_1 < 0 \\ \frac{z_1 - 0}{0.175 - 0} & 0 \leq z_1 \leq 0.175 \\ \frac{0.35 - z_1}{0.35 - 0.175} & 0.175 \leq z_1 \leq 0.35 \\ 0 & z_1 > 0.35 \end{cases} \tag{2}$$

$$\mu_{Medium}(z_1) = \begin{cases} 0 & z_1 < 0.35 \\ \frac{z_1 - 0.35}{0.495 - 0.35} & 0.35 \leq z_1 \leq 0.495 \\ \frac{0.64 - z_1}{0.64 - 0.495} & 0.495 \leq z_1 \leq 0.64 \\ 0 & z_1 > 0.64 \end{cases} \tag{3}$$

$$\mu_{High}(z_1) = \begin{cases} 0 & z_1 < 0.64 \\ \frac{z_1 - 0.64}{0.82 - 0.64} & 0.64 \leq z_1 \leq 0.82 \\ \frac{1.00 - z_1}{1.00 - 0.82} & 0.82 \leq z_1 \leq 1.00 \\ 0 & z_1 > 1.00 \end{cases} \quad (4)$$

Where z_1 = input value of aggression variable for which to calculate membership values of different fuzzy sets. Therefore, from equation (1) the values of α^*, β^*, and γ^* for aggression fuzzy sets are shown in Table 17.1.

Table 17.1 Linguistic ranges for input variable aggression.

Fuzzy sets of Aggression input variable	Triangular crisp values for the input aggression
Low	$[0, 0.175, 0.35]$
Medium	$[0.35, 0.495, 0.64]$
High	$[0.64, 0.82, 1.00]$

Similarly, the membership function's equations, and their corresponding triangular crisp values of remaining inputs and output variables can be calculated.

The fuzzy rule base is defined in the next step after analyzing the whole system. "Fuzzy rule base consists of many fuzzy rules, such as *"if-then"* which represents the relation of the system between input and output variables". In this proposed system, the total number of rule-base is 243 with five input and one output variable. Some of them are shown below in Figure 17.9.

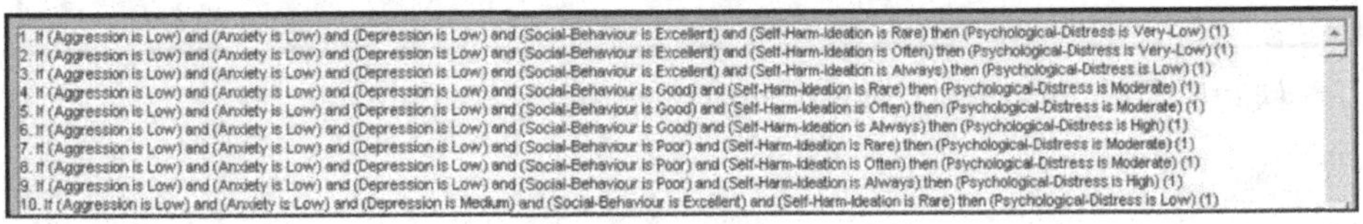

1. If (Aggression is Low) and (Anxiety is Low) and (Depression is Low) and (Social-Behaviour is Excellent) and (Self-Harm-Ideation is Rare) then (Psychological-Distress is Very-Low) (1)
2. If (Aggression is Low) and (Anxiety is Low) and (Depression is Low) and (Social-Behaviour is Excellent) and (Self-Harm-Ideation is Often) then (Psychological-Distress is Very-Low) (1)
3. If (Aggression is Low) and (Anxiety is Low) and (Depression is Low) and (Social-Behaviour is Excellent) and (Self-Harm-Ideation is Always) then (Psychological-Distress is Low) (1)
4. If (Aggression is Low) and (Anxiety is Low) and (Depression is Low) and (Social-Behaviour is Good) and (Self-Harm-Ideation is Rare) then (Psychological-Distress is Moderate) (1)
5. If (Aggression is Low) and (Anxiety is Low) and (Depression is Low) and (Social-Behaviour is Good) and (Self-Harm-Ideation is Often) then (Psychological-Distress is Moderate) (1)
6. If (Aggression is Low) and (Anxiety is Low) and (Depression is Low) and (Social-Behaviour is Good) and (Self-Harm-Ideation is Always) then (Psychological-Distress is High) (1)
7. If (Aggression is Low) and (Anxiety is Low) and (Depression is Low) and (Social-Behaviour is Poor) and (Self-Harm-Ideation is Rare) then (Psychological-Distress is Moderate) (1)
8. If (Aggression is Low) and (Anxiety is Low) and (Depression is Low) and (Social-Behaviour is Poor) and (Self-Harm-Ideation is Often) then (Psychological-Distress is Moderate) (1)
9. If (Aggression is Low) and (Anxiety is Low) and (Depression is Low) and (Social-Behaviour is Poor) and (Self-Harm-Ideation is Always) then (Psychological-Distress is High) (1)
10. If (Aggression is Low) and (Anxiety is Low) and (Depression is Medium) and (Social-Behaviour is Excellent) and (Self-Harm-Ideation is Rare) then (Psychological-Distress is Low) (1)

Figure 17.9 Rule base for system analysis

After rule base formation, defuzzification is performed using the centroid method, which is given below, through which the status of psychological distress of college students can easily be checked by the rule viewer.

$$z_{Centroid} = \frac{\sum_{i=0}^{n} \mu(z_i).(z_i)}{\sum_{i=0}^{n} \mu(z_i)} \quad (5)$$

In this process, Figure 17.10 having input and output variables shows a case's status, if input variables – aggression is 77.8% (high), anxiety is 84.6% (high), depression is 84% (high), social behavior is 87.7% (poor) and self-harm ideation is 10.8% (rare) then that student is experiencing 92.5% (very high) psychological distress.

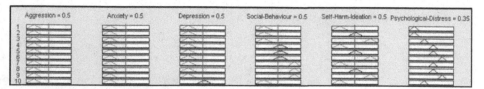

Figure 17.10 Psychological distress analysis with rule viewer

Here we can also present a 3-D view of two input variables with output variable with the help of a surface viewer. Some of them are shown by Figures 17.11–17.14.

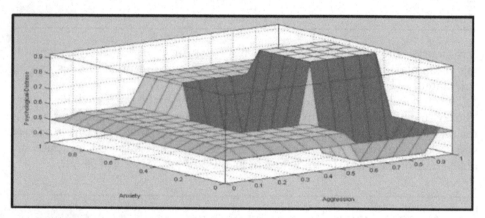

Figure 17.11 3-D surface view of anxiety, aggression & output

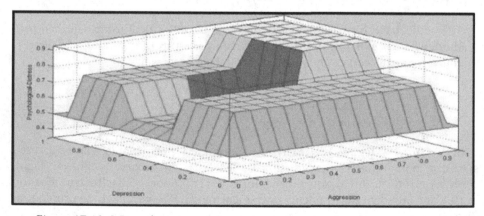

Figure 17.12 3-D surface view of depression, aggression & output

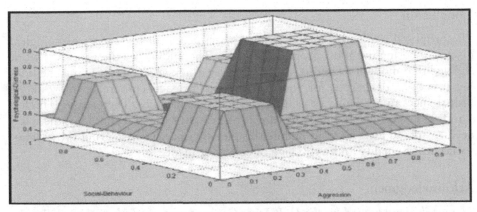

Figure 17.13 3-D surface view of social behavior, aggression & output

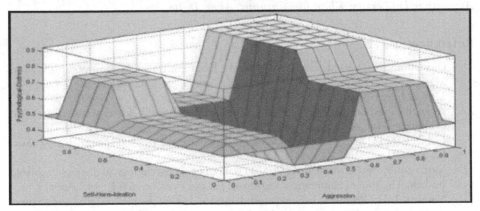

Figure 17.14 3-D surface view of self-harm ideation, aggression & output

Result

The result of the proposed work has indicated that a significant percentage of college students who participated in the offline survey are psychologically distressed due to excessive use of modern technology. It shows that 42% of respondents suffer from high level of psychological distress, 31% from very high level, 13% from moderate, 9% from low level, and 5% from very low level of psychological distress. The graphical presentation of this result is given below.

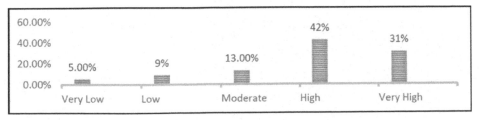

Figure 17.15 Percentage of college students going through different levels of psychological distress

Conclusion

The proposed work uses the fuzzy logic system concept to assess psychological distress by using five most prominent factors. The findings of the proposed work showed that modern technology provides very quick access and various facilities but it is quite dangerous for young users as its impacts can create unavoidable disturbances in academic as well as practical performance. Therefore, by understanding the terrible and negative effects of excessive use of technology, there is a need to spread awareness to handle this global concern considering the psycho-social health quality and well-being of college students.

Acknowledgment

The work is supported by the U. P. State Government research project under the supervision of Prof. Sanjeev Kumar, Department of Mathematics, Dr. Bhimrao Ambedkar University, Khandari Campus, Agra, India.

References

1. Diomidous, M., Chardalias, K., Magita, A., Koutonias, P., Panagiotopoulou, P., and Mantas, J. (2016). Social and psychological effects of internet use. *ACTA Inform. Med.*, 24(1), 66–69.
2. Hoge, E., Bickham, D., and Cantor, J. (2017). Digital media, anxiety, and depression in children. *Am. Acad. Pediat.*, 140(2), 76–80.
3. Joshi, A. (2021). Covid-19 pandemic in India: through psycho-social lens. *J. Soc. Econom. Dev.*, 23(2), 414–437.
4. Morrison, R. and Connor, R. C. O. (2005). Predicting psychological distress in college student: The role of rumination and stress. *J. Clin. Psychol.*, 61(4), 447–460.
5. Nguyen, T. H., Rahman, F. F., Peterou, J., and Wong, W. K. (2020). Study of depression, anxiety, and social media addiction among undergraduate students. *J. Manag. Inform. Dec. Sci.*, 23(4), 257–276.
6. Nsereko, N. D., Musisi, S., Janet, N., and Ssekiwu, D. (2014). Psychosocial problems and development of psychopathology among Ugandan university students. *Int. J. Res. Stud. Psychol.*, 1(1), 1–15.
7. Pachiyappan, T., Mark, P., Kumar, K. V., Venugopal, R., Palanisamy, B., and Jilumudi, D. (2021). Effects of excessive usage of electronic gadgets during covid-19 lockdown on the health of college students. *Asian J. Pharma. Res. Health Care*, 13(2), 139–145.
8. Zadeh, L. A. (1965). Fuzzy sets. *Inform. Con.*, 8(3), 338–353.

18 A comparative study prediction of heart disease by using regression modeling: Evidence from NFHS-5

Divya[1,a] and Prof. Vineeta Singh[2,b]

[1]Research Scholar, Department of Statistics, Institute of Social Science, Dr. Bhimrao Ambedkar University, Agra, Uttar Pradesh, India

[2]Professor, Department of Statistics, Institute of Social Science, Dr. Bhimrao Ambedkar University, Agra, Uttar Pradesh, India

Abstract

The heart is a very important organ of human beings. Life is completely dependent on the dexterous working of the heart. Cardiovascular diseases are the major and most dangerous diseases in the world. According to the WHO report 17.9 million people are suffering from heart disease. So, it is crucial to detect and diagnose heart disease early. Many statistical, mathematical, and computer science data mining techniques are useful in predicting heart disease at the early stage. So, we aim to predict heart disease in the early stage. In this study, we propose an efficient logistic regression model of data mining for the prediction of heart disease. For this study, we used a benchmark dataset the National Family Health Survey-5 (NFHS-5) 2019–21 dataset of Uttar Pradesh, India. We applied logistic regression algorithm on the women and men datasets individually with different attributes for the prediction of disease. We obtained the classification accuracy for women at 99.5% and for men at 99.6% by the models. The researchers propose an immediate and efficient algorithm for diagnoses of heart disease for men as well as women.

Keywords: Heart disease, data mining, logistic regression, NFHS

1. Introduction

Non-communicable diseases often encompass conditions such as cardiovascular disease (CVD), different forms of cancer, chronic respiratory disorders, diabetes, and similar ailments. These are projected to be the cause of approximately 60% of all fatalities. The biggest cause of death worldwide is cardiovascular disease (CVD) at 17.9%, which includes ischemic heart disease and cerebrovascular diseases like stroke. According to projections, there would be 4.77 million CVD-related fatalities annually in India by 2020, up from 2.26 million in 1990. Across the last few decades, assessments of coronary heart disease in India's rural and urban communities have ranged from 1.6% to 7.4% and 1% to 13.2%, respectively. Global fatalities are predominantly attributed to cardiovascular diseases; various factors are responsible for heart disease such as age, obesity, hypertension, high blood pressure healthy lifestyles, etc. It is feasible to recognize a variety of symptoms by observing physical signs like dizziness, shortness of breath, chest pain, etc. so, it is essential to discover heart disease at an early otherwise its

[a]divyascholar7999@gmail.com, [b]vineetasanjaisingh@gmail.com

complications can influence a person's life. The signs of a woman having heart disease are much less those of men. Women may feel pain in the center of the chest, high blood pressure, squeezing, nausea, etc. Men are faced with symptoms of heart attack, such as stress, discomfort, chest pain, etc. They both also feel pain in the other body parts, such as the jaw, neck, arms, and also sweating, and shortness of breath. Many data mining algorithms are useful for the prediction of heart disease. A smart and efficient technique is demanded to decrease the high death rate caused by heart disease. In this study, we used the smart statistical technique used for data mining logistic regression to detect heart disease as soon as possible. Our study comprises 5 sections. Introduction, Recent researches are related to heart diseases, The proposed system, Experimental results and finally conclusion.

2. Literature survey

T. et al. (2013) detected heart disease by using logistic regression, decision tree and SVM models with rule-based algorithm and attained the multiple rule algorithm as the best performer. Khanna et al. (2015) predicted heart disease by using logistic regression, SVM and neural networks algorithms and got accuracy of 88.2%, 87%, and 85% by the models. Abdar et al. (2015) used C5.0, neural network, support vector machine, K-Neighborhood, and logistic regression to predict heart disease and got the highest accuracy with 93.02% from the C5.0 algorithm. Jindal et al. (2021) used machine learning algorithms such as logistic regression, random forest and KNN to predict heart disease and got an accuracy of the model is 87.5%. In their study, Karthick et al. (2022) employed logistic regression, SVM, Gaussian Naïve Bayes, Light GBM, XG Boost, and random forest algorithms to forecast heart disease. The achieved accuracy rates across different models were as follows: 80.32%, 80.32%, 78.68%, 77.04%, 73.77%, and 88.5%, respectively. Yilmaz and Yağin (2021) early detected the heart disease using logistic regression, SVM and random forest and got an accuracy of 86.1%, 89.7% and 92.9%.

3. Proposed methodology

Today, many hospitals use healthcare information systems to manage patient data since these systems have vast amounts of data from which it is possible to extract hidden information for smart medical diagnosis. This research aims to construct a proficient model for predicting heart disease, facilitating early diagnosis. Figure 18.1 shows the proposed system architecture. The data used in this model is the National Family Heath Survey-5 (NFHS-5) 2019–21 dataset (https://dhsprogram.com/data/dataset_admin/index.cfm) of Uttar Pradesh India. In our study, the target variable or dependent variable is heart disease and the independent variables are alcohol, diabetes, hypertension, sex, smoking, chewing tobacco, etc., as described in Table 18.1. These datasets are performed in IBM SPSS Version.29. In the data pre-processing step, we imputed our missing data by predicting mean matching (PMM) and multiple imputation methods. PMM is used for filling scale attributes and the multiple imputation method is used for filling categorical

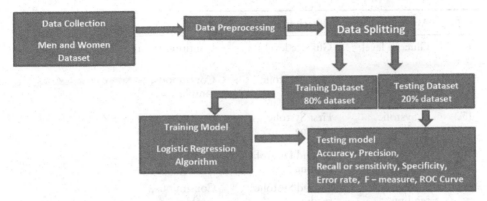

Figure 18.1 Architecture of proposed system

Table 18.1 Various attributes used for heart disease prediction.

S. No.	Attributes	Description	Value	Men	Women
1.	Alcohol	Currently drink alcohol	Discrete, 0= No, 1= Yes	✓	✓
2.	Diabetes	Currently have diabetes	Discrete, 0= No, 1= Yes	✓	✓
3.	Hypertension	Currently have hypertension	Discrete, 0= No, 1= Yes	✓	✓
4.	Respiratory disease	Currently have respiratory disease	Discrete, 0= No, 1= Yes	✓	✓
5.	Thyroid disorder	Currently have thyroid disease	Discrete, 0= No, 1= Yes	✓	✓
6.	Cancer	Currently have cancer	Discrete, 0= No, 1= Yes	✓	✓
7.	Kidney disorder	Currently have kidney disorder	Discrete, 0= No, 1= Yes	✓	✓
8.	Residence place	Type of place of residence	Discrete, 1= Urban, 2= Rural	✓	✓
9.	Pregnant	Currently pregnant	Discrete, 0= No, 1= Yes	✗	✓
10.	Breastfeeding	Currently breastfeeding	Discrete, 0= No, 1= Yes	✗	✓
11.	Smoking	Currently smoking	Discrete, 0= No, 1= Yes	✓	✓
12.	Chews tobacco	Currently chews tobacco	Discrete, 0= No, 1= Yes	✓	✓
13.	Arm circumference	Arm circumference	Continuous, mm	✓	✓
14.	Age	Age in years	Continuous, Years	✓	✓

S. No.	Attributes	Description	Value	Men	Women
15.	Glucose level	Glucose level	Continuous, mg/dL	✓	✓
16.	1_Diastolic reading	First Diastolic reading	Continuous, mmHg	✓	✓
17.	1_Systolic reading	First Systolic reading	Continuous, mmHg	✓	✓
18.	2_Diastolic reading	Second Diastolic reading	Continuous, mmHg	✓	✓
19.	2_Systolic reading	Second Systolic reading	Continuous, mmHg	✓	✓
20.	3_Diastolic reading	Third Diastolic reading	Continuous, mmHg	✓	✓
21.	3_Systolic reading	Third Systolic reading	Continuous, mmHg	✓	✓
22.	Hemo level	Hemoglobin level	Continuous, g/L	✗	✓
23.	BMI	Body mass index	Continuous, kg/m²	✗	✓
24.	RI	Rohrer's index	Continuous, kg/m³	✗	✓
25.	WC	Waist circumference	Continuous, mm	✓	✓
26.	HC	Hip circumference	Continuous, mm	✓	✓
27.	Chews khaini	Chews khaini	Discrete, 0= No, 1= Yes	✓	✗
28.	Working	Currently working	Discrete, 0= No, 1= Yes	✓	✗
29.	Heart disease	Currently have heart disease	Discrete, 0= No, 1= Yes	✓	✓

attributes. In the data splitting step, we split datasets into two groups training dataset (80%) for the training model and the testing dataset (20%) for the model's performance. For the training and testing of the model, we used a logistic regression algorithm and to check the performance of the model and comparison of models, we used accuracy, precision, recall or sensitivity, specificity, error rate, F-measure, and receiver operating curve (ROC) curve.

4. Experimental results and discussion

Firstly, we describe the analysis of the training dataset of women and men.

Block 1: Method = Enter In this phase, we observed the overall test of the model. In the logistic regression model first, we checked the goodness of fit test by the Omnibus test. In our model, the test is statistically significant, with the value of $\chi^2 (27) = 1101.516$, $p<0.05$ for women and $\chi^2 (24) = 155.402$, $p< 0.05$ for men. Then we included the Pseudo R^2 and the -2log likelihood which is the minimization criteria used by SPSS. From this, we computed the total explained variation from the Cox & Snell R^2 and Nagelkerke R^2 values. Hence, the total variation in the dependent variable based on our models ranges from 1.5% to 25.3% for women and 1.6% to 23.5% for men which indicated the model is a good fit. And last we saw the confusion matrix (Tables 18.2 and 18.3) from these tables we calculated the classification accuracy, precision, recall or sensitivity, specificity, error rate, and F-measure of the training dataset are **99.6%, 90.1%, 15.7%, 100%, 0.00401%, 27%** for women and **99.5%, 93%, 10.9%, 100%, 0.0052%, 19.6%** for men.

Table 18.2 For women.

Currently has heart disease	Predicted		Percentage correct
	Currently has heart disease		
	No	Yes	
No	74227	6	100.0
Yes	295	55	15.7
Overall percentage			99.6

Table 18.3 For men.

Currently has heart disease	Predicted		Percentage correct
	Currently has heart disease		
	No	Yes	
No	9612	2	100.0
Yes	49	6	10.9
Overall percentage			99.5

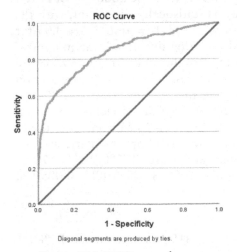

Figure 18.2 ROC curve for women

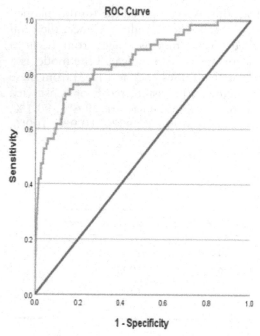

Figure 18.3 ROC curve for men

Confusion matrix

The area under the curve of the training dataset is **83%** for women and **84%** for men (Figures 18.2 and 18.3).

Now, we describe the analysis of the testing dataset of women and men.

Block 1: Method = Enter In this phase, we observed the overall test of the model. In the logistic regression model first, we checked the goodness of fit test by the Omnibus test. In our model, the test is statistically significant, with the value of $\chi^2(1) = 151.973$, $p<0.05$ for women, and $\chi^2(1) = 46.659$, $p< 0.05$ for men. Then we included the Pseudo R^2 and the -2log likelihood, the minimization criteria used by SPSS. From this, we computed the total explained variation from the Cox & Snell R^2 and Nagelkerke R^2 values. Hence, the total variation in the dependent variable based on our models ranges from 8% to 14.8% for women and 0.7% to 11.4% for men which indicated the model is a good fit. And last we saw the confusion matrix (Tables 18.4 and 18.5) from these tables we calculated the classification accuracy, precision, recall or sensitivity, specificity, error rate, and F-measure of the testing dataset are **99.5%, 81%, 17.1% 100%, 0.0037%, 28%** for women, and **99.6%, 50%, 8.3%, 100%, 0.0050%, 14.2%** for men.

Confusion matrix

The area under the curve of the testing dataset is **81%** for women and **75%** for men (Figures 18.4 and 18.5, Table 18.6).

Table 18.4 For women

	Predicted		
Currently has heart disease	Currently has heart disease	Percentage correct	
	No	Yes	
No	18458	1	100.0
Yes	68	14	17.1
Overall percentage			99.6

Table 18.5 For men

	Predicted		
Currently has heart disease	Currently has heart disease		Percentage correct
	No	Yes	
No	2361	1	100.0
Yes	11	1	8.3
Overall percentage			99.5

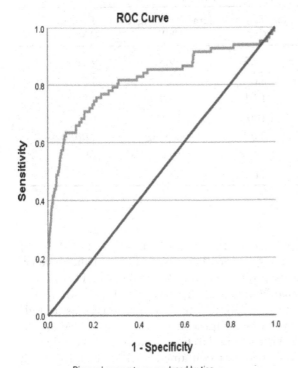

Diagonal segments are produced by ties.

Figure 18.4 ROC curve for women

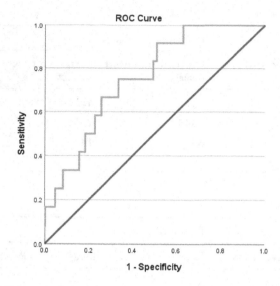

Figure 18.5 ROC curve for men

Table 18.6 Comparison of results of logistic regression models.

Evaluation methods	Women		Men	
	Training data (%)	Testing data (%)	Training data (%)	Testing data (%)
Accuracy	99.6	99.5	99.5	99.6
Precision	90.1	81	93	50
Recall or sensitivity	15.7	17.1	10.9	8.3
Specificity	100	100	100	100
Error rate	0.00401	0.0037	0.0052	0.0050
F-measure	27	28	19.6	14.2
ROC curve	83	81	84	75

Conclusion

A heart disease prediction model was created through the utilization of the data mining technique known as logistic regression. This research establishes a reference point for current studies focused on predicting heart disease. The dataset utilized originates from the National Family Health Survey (NFHS-5) conducted during 2019–21, which is not only carefully curated but also serves as a credible standard for research. This paper has comprehensively outlined the contrast between logistic regression models when applied to the prediction of heart disease in both women and men. The women's data contained 93,124 instances with 5,912 attributes and the men's data contain 12,043 instances with 861 attributes based on previous studies we only took useful 27 attributes for women and 24 attributes for men for predicting heart disease and these datasets were performed on IBM SPSS Version.29. Based on the investigation method logistic regression achieved the predicting model accuracy of 99.5% for women and 99.6% for men. This study helps society people, and doctors, to analyze and recognize their health, and patient health status. This will also predict other diseases and based on that we can diagnose diseases. For future work, we can expand our work by using hybrid classification techniques with other tools for the prediction and diagnosis of other deadly diseases.

References

1.   Abdar, M., Kalhori, S. R. N., Sutikno, T., Subroto, I. M. I., and Arji, G. (2015). Comparing the performance of data mining algorithms in prediction heart diseases. *Int. J. Elec. Comp. Engg.* 5(6), 1569–1576. https://doi.org/10.11591/ijece.

2.   Jindal, H., Agrawal, S., Khera, R., Jain, R., and Nagrath, P. (2021). Heart disease prediction using machine learning algorithms. *IOP Conf. Series: Mat. Sci. Engg.*, 1022(1), https://doi.org/10.1088/1757-899x/1022/1/012072.

3.   Karthick, K., Aruna, S. K., Samikannu, R., Kuppusamy, R., Teekaraman, Y., and Thelkar, A. R. (2022). Implementation of a heart disease risk prediction model using machine learning. *Comput. Math. Methods Med.*, 2022. https://doi.org/10.1155/2022/6517716.

4.   Khanna, D., Sahu, R., Baths, V., and Deshpande, B. (2015). Comparative study of classification techniques (SVM, logistic regression and neural networks) to predict the prevalence of heart disease. *Int. J. Machine Learn. Comput.*, 5(5), 544. https://doi.org/10.7763/ijmlc.

5.   T., M., Mukherji, D., Padalia, N., and Naidu, A. (2013). A heart disease prediction model using SVM-decision trees-logistic regression (SDL). *Int. J. Comp. Appl.*, 68(16), 11–15. https://doi.org/10.5120/11662-7250.

6.   Yilmaz, R. and Yağin, F. H. (2021). Early detection of coronary heart disease based on machine learning methods. *Med. Records.*, https://doi.org/10.37990/medr.1011924.

7.   Data Source (https://dhsprogram.com/data/dataset_admin/index.cfm.

19 Exponentially varying lead time crashing cost stochastic inventory model with provision of saving money on backorders

Vineet Kumar[a] and Sharon Moses[b]

Research Scholar, Associate Professor, Department of Mathematics, St. John's College, Agra, Dr. Bhimrao Ambedkar University, Agra, India

Abstract

An exponentially distributed lead time crash inventory model with safety element and the price reduction for backorders is analyzed to estimate the cost of reducing lead time using an exponential distribution. This means that decreasing lead time can be achieved by raising the cost of crashing. The goal of this study is to determine the lowest possible all-in price for a specified period of lead time. We describe the method and we used to solve the problem using MATLAB. A numerical example is also provided to demonstrate the underlying mathematical structure. the results are verified graphically.

Keywords: Inventory, lead time, crashing cost

Introduction

Prologue – Stock administration and control is worried about the procurement and capacity of materials important to support an organization's exercises. Reducing waiting times is another important aspect of integrated inventory management. Lead time is a key factor in inventory management, which has been the subject of numerous books and articles. Managing inventory and the supply chain effectively requires a thorough understanding of lead time. An item's "lead time" is how much time it requires from the investment once a request is made until the item becomes prepared for procurement. Most studies on inventory management use either a deterministic or probabilistic model in which lead time is a constant that cannot be altered.

Two studies, one by Vijayashree and Uthayakumar (2016) and another by Priyan and Uthayakumar (2015), hypothesized that the crashing costs would increase exponentially with advance notice. Using a two-tier supply chain model, Kang and Kim (2010) emphasized the importance of coordinating stockpiling and logistics management. According to Vijayashree and Uthayakumar (2016), the crashing cost as a function of lead time is exponential. To get to the best answer, they came up with a technique for finding solutions. The supply chain model developed by Kim and Sarkar (2017) accounts for stochastic demand, a trade credit policy, improved product quality, lower supplier setup costs, and a variable backorder rate. In their study, Rekha and Pavithra (2018) discussed an inventory model for the food processing and distribution business in which the

[a]vineetshukla473@gmail.com, [b]sharonmoses38@gmail.com

cost of holding inventory drops exponentially with the length of time it takes to replenish it. Authoritative manageability practices, methodologies, and capacities develop over the course of time, as exhibited by Chowdhury et al. (2019). This is in tandem with shifts in the priorities of stakeholders towards sustainability requirements. With the help of a Weibull function, Krishnaraj (2020) created inventory models in which the cost of a crash in lead time is modeled. In this approach, the lead time crashing cost is represented as a Weibull distribution. In order to maximize profits while still providing excellent customer service. Sarkar et al. (2021) advocated a long-term online-to-offline selling approach. To research the drawn-out conduct of the arrangements of constant audit (s, S) stock strategy, Islam et al. (2022) fostered a model for level-subordinate short-lived items with positive help time on a Jackson queuing network.

Definitions and terms glossary

d = Demand
q = Order quantity
o = Price paid for each order
I = Price of inventory
H = Cost of holding
R^P = Reorder point
l^t = Lead time in weeks
T^c = Total cost
A = Decreasing the cost and lead time
θ_0 = Profit at the margin (increment to profit)
θ_x = Price reductions for backorders
μl^t = Mean during lead time l^t
$\sigma\sqrt{l^t}$ = Standard deviation during lead time l^t

Formulation of model

All costs are calculated as per the preceding assumptions and notations.

$$\text{Total Cost} = T^C(q, l^t) = \text{Ordering} + \text{Holding} + \text{shortage} + \text{lead time crashing cost} \tag{1}$$

$$\text{Annualized ordering expenses are estimated as } \frac{O \times d}{q} \tag{2}$$

$$\text{Just for the sake of argument, let's say } R^P = d \times l^t + S_f \times \sigma\sqrt{l^t} \tag{3}$$

Where S_f = safety factor
Allowing for a linear decline over the course of the cycle

$$I = \frac{q}{2} + R^P - d \times l^t \tag{4}$$

By plugging the R^P values from equation (3) into equation (4), we obtain

$$I = \frac{q}{2} + d \times l^t + S_f \times \sigma\sqrt{l^t} - d \times l^t = \frac{q}{2} + S_f \times \sigma\sqrt{l^t} \tag{5}$$

lead time crashing cost $= A = \frac{d}{q} \times R^P(l^t)$ (6)

$$R^P(l^t) = \begin{cases} 0 & l^t = l_2 \\ \frac{K}{e^{l^t}} & l_1 \leq l^t < l_2 \end{cases} \tag{7}$$

The backorder ratio γ is not constant; rather, it fluctuates according to the percentage of price reduction that the supplier offers for backorders per unit θ_x.

Thus $\gamma = \frac{\gamma\theta_x}{\theta_0}, \ 0 \leq \gamma \leq 1, 0 \leq \theta_x \leq \theta_0$ (8)

It is possible to represent the anticipated deficiency in supply at the end of a cycle as

$$\frac{d}{q}[\theta_x\gamma - \theta_0(1 - \gamma)] \tag{9}$$

The solution to equation (1) is given below

$$T^C(q, l^t) = \frac{O \times d}{q} + H \times I + \frac{d}{q}[\theta_x\gamma - \theta_0(1 - \gamma)] + \frac{d}{q} \times R^P(l^t) \tag{10}$$

When we plug the values of $R^P(l^t)$ into equation (10) that we got from equation (7), we get the following:

From equations (1),

$$T^C(q, l^t) = \frac{O \times d}{q} + H \times \left(\frac{q}{2} + S_f \times \sigma\sqrt{l^t}\right) + \frac{d}{q}[\theta_x\gamma - \theta_0(1 - \gamma)] + \frac{d}{q} \times e^{\frac{K}{l^t}} \tag{11}$$

Solution of the problem

Obtaining a solution for equation (11) by performing partial differentiation with regard to q and l^t correspondingly

$$\frac{\partial T^C}{\partial q} = -\frac{O \times d}{q^2} + \frac{H}{2} - \frac{d}{q^2} \times [\theta_x\gamma - \theta_0(1 - \gamma)] - \frac{d}{q^2} \times e^{\frac{K}{l^t}} = 0 \tag{12}$$

$$\frac{\partial T^C}{\partial l^t} = \frac{1}{2\sqrt{l^t}} HS_f\sigma - \frac{d}{q}e^{\frac{K}{l^t}}\left[\frac{K}{(l^t)^2}\right] = 0 \tag{13}$$

We get the following result from equation 12

$$-\frac{O \times d}{q^2} + \frac{H}{2} - \frac{d}{q^2} \times [\theta_x \gamma - \theta_0(1-\gamma)] - \frac{d}{q^2} \times e^{\frac{K}{l^t}} = 0$$

$$\frac{H}{2}q^2 = d\left[O + [\theta_x \gamma - \theta_0(1-\gamma)] + e^{\frac{K}{l^t}}\right]$$

$$q^2 = \frac{2d}{H}\left[O + [\theta_x \gamma - \theta_0(1-\gamma)] + e^{\frac{K}{l^t}}\right] \qquad (14)$$

$$q = \sqrt{\frac{2d}{H}\left[O + [\theta_x \gamma - \theta_0(1-\gamma)] + e^{\frac{K}{l^t}}\right]}$$

We get the following from equation (13)

$$\frac{1}{2\sqrt{l^t}}HS_f\sigma - \frac{d}{q}e^{\frac{K}{l^t}}\left[\frac{K}{(l^t)^2}\right] = 0$$

$$\frac{1}{2}HS_f\sigma \times (l^t)^{3/2} = \frac{d}{q}Ke^{\frac{K}{l^t}} \qquad (15)$$

$$(l^t)^{3/2} = \frac{2d}{q}Ke^{\frac{K}{l^t}}\frac{1}{HS_f\sigma}$$

$$l^t = \left[\frac{2d}{q}Ke^{\frac{K}{l^t}}\frac{1}{HS_f\sigma}\right]^{2/3}$$

Numerical illustration and discussion

In order to demonstrate the efficacy of the aforementioned approach to solving the problem, a numerical example is presented here.

Supplier 1: Allow us to consider the stock framework with the accompanying information = 700 *unit/year*, $0 = 250$ Rs. per order, $H = 23$, $\sigma = 8$ *units/week*, $S_f = 2.5$ and lead time crashing cost and furthermore backorder cost rebate is given as follows; $\theta_0 = 150$ *per unit*, $\theta_x = 50$ Rs. what's more, $\gamma = 0.85$.

Supplier 2: Allow us to consider the stock framework with the accompanying information $d = 700$ *unit/year*, $0 = 275$ Rs. per order, $H = 25$, $\sigma = 8$ *units/week*, $S_f = 2.8$ and lead time crashing cost and furthermore backorder cost rebate is given as follows; $\theta_0 = 150$ *per unit*, $\theta_x = 60$ Rs. and $\gamma = 0.88$.

Supplier 3: Allow us to consider the stock framework with the accompanying information = 700 *unit/year*, $0 = 300$ Rs. per order, $H = 28$, $\sigma = 8$ *units/week*, $S_f = 3.0$ and lead time crashing cost and furthermore backorder cost rebate is given as follows; $\theta_0 = 150$ *per unit*, $\theta_x = 70$ Rs. and $\gamma = 0.92$.

The cost of crashing the lead time is listed down below:

$$R^P(l^t) = \begin{cases} 0 & l^t = 8 \\ \frac{K}{e^{l^t}} & 1 \le l^t < 7 \end{cases}, K = 7$$

The computational results, which were obtained by applying the solution process are presented in Table 19.1. The ideal answer, according to Table 19.1, for supplier 1 is to have a lead time of 2 weeks, an order quantity of 128.3273 units, and an integrated total cost of 3,779.7. This information can be read-off as follows: The ideal option, based on Table 19.1, can be read off as the following for

Table 19.1 of Exponentially varying lead time crashing cost stochastic inventory model with provision of saving money on backorders

l^t	Supplier-1		Supplier-2		Supplier-3	
	q	T^C	q	T^C	q	T^C
1	132.4063	9207.7	135.814	9506.6	136.8087	10001
2	**128.3273**	**3779.7**	**131.8405**	**4260.7**	**132.865**	**4842**
3	128.2384	3801.6	131.754	4317.6	132.7792	4935
4	128.2207	3900	131.7367	4443.4	132.762	5091
5	128.214	3999.3	131.7302	4566.6	132.7556	5241
6	128.2107	4092.9	131.727	4681.7	132.7524	5380
7	128.2088	4180.4	131.7252	4788.9	132.7506	5509
8	128.2076	4262.7	131.7239	4889.5	132.7494	5630

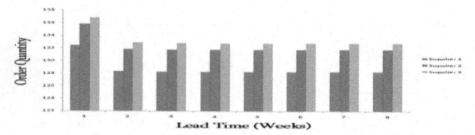

Graph 19.1 Order Quantity Vs Lead Time

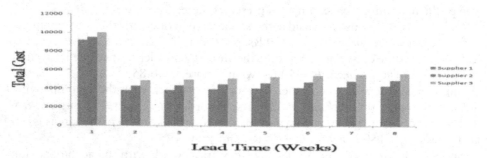

Graph 19.2 Total Cost Vs Lead Time

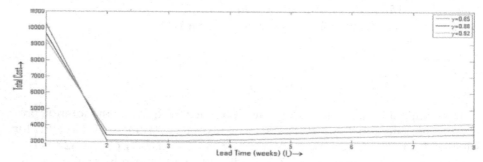

Graph 19.3 Totla Cost Vs Lead Time for different values of Backorder Price Discount

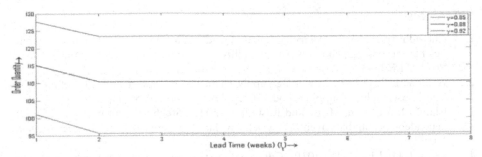

Graph 19.4 Order Quantity Vs Lead Time for different values of Backorder Price Discount

supplier 2: lead time of two weeks, order quantity of 131.8405 units, and the associated integrated total cost of 4266.7. Similarly, the ideal solution for supplier 3 may be read-off of Table 19.1 as having a lead time of 2 weeks, an order quantity of 132.865 units, and a matching integrated total cost of 4842. This can be read-off. In light of the information presented in the preceding model, we are able to draw the conclusion that supplier 1 is the most advantageous of the three suppliers because of the 2-week lead time that they offer. This may be seen in the bar Graphs 19.1 and 19.2 as well (2). We have showed the fluctuation in total cost and order quantity with lead time for various values of back price discount in Graphs 19.3 and 19.4, respectively. These graphs can be found below. It has been observed that larger back price discounts result in greater order quantities.

Concluding remarks

Lead time is a trendy practice in today's automated inventory and supply chain management systems. Any method of inventory management worth its salt will take lead time into account. The time that elapses between the beginning of an activity and its conclusion is called its lead time. Lead time was previously viewed as an uncontrollable constant in conventional inventory models. But in actuality, you can shorten the lead time by paying more for the crash. In many real-world contexts, lead time can be shortened for a tiny additional crash cost. That is to say, waiting times can be controlled. So, we present stock-keeping models that think about the number of orders and how long it will take to fulfill them. The paper's central premise is that the per-order crashing cost $R^p(l^t)$ increases at an exponential rate as the lag time l^t increases. By means of a numerical example, the constructed model is demonstrated to be effective. This study makes a unique contribution by creating a mathematical model and a reliable method for solving it. Important challenge associated with the suggested models are also highlighted via a numerical example. The supplier and the merchant both want to maximize their mutual profit, hence the integrated inventory model is gaining traction. Here, we create the technique for solving the problem in MATLAB and determining the best possible answer.

References

1. Chowdhury, M. M. H., Agarwal, R., and Quaddus, M. A. (2019). Dynamic capabilities for meeting stakeholders' sustainability requirements in supply chain. *J. Clean. Prod.*, 215, 34–35.
2. Ganeshan, R. (1999). Managing supply chain inventories: a multiple retailer, one warehouse, multiple supplier model. *Int. J. Prod. Econom.*, 59(1), 341–354.
3. Islam, M. A., Islam, M. E., and Rashid, A. (2022). Stochastic optimization of level-dependent perishable inventory system by Jackson network. *Ain Shams Engg. J.*, 101935.
4. Kang, J. H., Kim, Y. D. (2010). Coordination of inventory and transportation managements in a two-level supply chain. *Int. J. Prod. Econom.*, 123(1), 137–145.
5. Kim, S. J. and Sarkar, B. (2017). Supply chain model with stochastic lead time, trade-credit financing, and transportation discounts. *Math. Prob. Engg.*, 14. Article ID 6465912.
6. Krishnaraj, R. B. (2020). Modeling for inventory involving lead time crashing cost as a weibull distribution with backorder price discount. *J. Interdis. Cycle Res.*, XII(VI), 379–383.
7. Priyan S. and Uthayakumar R. (2015). Continuous review inventory model with controllable lead time, lost sales rate and order processing cost when the received quantity is uncertain. *J. Manufac. Sys.*, 34, 23–33.
8. Rekha, S. and Pavithra, P. (2018). Solving an inventory models involving lead time crashing cost as an exponential function in food processing and distribution industry using matlab. *Int. J. Sci. Res. Sci. Engg. Technol.*, 556–560.
9. Sarkar, B., Dey, B. K., Sarkar, M., and Al Arjani, A. (2021). A sustainable online-to-offline (o2o) retailing strategy for a supply chain management under controllable lead time and variable demand. *Sustainability*, 13, 1–26.
10. Vijayashree, M. and Uthayakumar, R. (2016). Inventory models involving lead time crashing cost as an exponential function. *Int. J. Manag. Value Supp. Chains*, 7(2), 29–39.
11. Vijayashree, M. and Uthayakumar, R. (2016). Two-echelon supply chain inventory model with controllable lead time. *Int. J. Sys. Assur. Engg. Manag.*, 7, 112–125.

20 Minimize various expenses arising during production-distribution planning under imprecise demands and production quantity by using fuzzy optimization approach

Saurabh Gupta[a] and Dr. Manju Sharma

Department of Mathematics, Dr. Bhimrao Ambedkar University, Agra, Uttar Pradesh, India

Abstract

In this paper, we have constructed a realistic model of production-distribution in which we have minimized the various expenses which arises during planning of production-distribution and this minimization is carried out by utilizing the fuzzy multi-objective linear programming approach proposed by Zimmermann and in this model the market demand, demand of raw materials, total demand of all the distributors and the production quantity are considered as fuzzy variable in nature. Now at the end of this paper we have discussed a numerical example for showing the applicability and usefulness of this research study and then finally we have presented the conclusion at the end of this research paper.

Keywords: Production-distribution, fuzzy variable, expenses, fuzzy multi-objective linear programming

1. Introduction

Ghanbari et al. (2020) investigated numerous problems of fuzzy linear programming using models and approaches to solutions. Goodarzian and Hosseini-Nasab (2021) presented a model of fuzzy multi-objective for an issue of production-distribution network design by utilized an algorithm of novel self-adoptive evolutionary. Ghanbarzadeh-Shams et al. (2022) proposed a model of fuzzy multi-objective programming along with chance constraints. Seif Barghi and Shirin Bayan (2023) minimized logistics cost, time of delivery and cost of CO_2 emission by constructed a model of two echelon multi-product and multi-period network of supply chain with production and centers for distribution.

2. Fuzzy multi-objective linear programming model and their assumptions

Assumptions
1. The production quantity, purchasing demand of raw material, total demands of all the distributors and market demand of any product is fuzzy variable in nature.

---

[a]saurabhgupta00979@gmail.com

2. Financial budget for purchasing the raw material, production of different types of products and different distributors is certain in this model.

3. In this model we have considered only one manufacturing unit which manufactured different types of products.

We have constructed Table 20.1 for notations used in a proposed model along with their explanation.

Table 20.1 of Minimize various expenses arising during production-distribution planning under imprecise demands and production quantity by using fuzzy optimization approach

	Notation	Explanation
Index set	k	Index for product which is going to be manufactured
	q	Index for distributor
	p	Index for supplier which delivered the raw materials
Decision variables	X_k	Amount of production of "k" type of product carried out by a manufacturing unit, for all k=1,2,3............. K
	Y_q	Amount of product transported to a distributor "q", for all q=1,2,3............. Q
	S_q	Amount of product transported from distributor "q" to a market, for all q=1,2....... Q
	A_p	Amount of raw material delivered by supplier "p", for all p=1,2,3......P
		Expenses arises during production planning of "k" type of product
For objectives	b_k	Unit production cost to produce "k" type of product by a manufacturing unit
	r_k	Cost of damaged raw material during per unit manufacturing of product "k"
	l_k	Electricity cost for manufacturing per unit of product "k"
	w_k	Worker cost during working hours for per unit manufacturing of product "k"
	h_k	Holding cost per unit of inventory of "k" type of manufactured product
	a_k	Holding cost per unit of inventory of raw material for "k" type of product
	u_k	Packaging cost of per unit of "k" type product for transportation
	o_k	Worker's overtime cost of per unit manufacturing of "k" type of product
	m_k	Worker cost for packaging of per unit of manufactured product "k"
		Expenses arises during distribution planning
	v_q	Total vehicle maintenance expenses of distributor "q" in one month

	Notation	Explanation
	d_q	Cost of packaging of per unit product in distributor "q"
	c_q	Fuel consumption cost during transporting product from distributor "q" to a market in one shift
	f_q	Fuel consumption cost during transporting product to a distributor "q" in one shift
	e_q	Holding cost per unit of inventory of finished product in distributor "q"
	n_q	Cost allotted to one driver for transporting product to a distributor "q"
	t_q	Cost allotted to one driver for transporting product from distributor "q" to a market
	j_q	Cost of amount of product damaged during delivery to a distributor "q" in one shift
	g_q	Cost of the amount of product damaged during transportation from distributor "q" to a market in one shift
For constraints	B^{max}	Maximum purchasing budget of raw material
	C_p	Per unit cost of raw material provided by supplier "p" to a manufacturing unit
	$\tilde{R}$	Fuzzy total requirement of all the raw materials for manufacturing
	$\tilde{F}$	Fuzzy total demand of a market
	$\tilde{J}$	Fuzzy total demand of all the distributors
	D_q^{max}	Maximum financial budget of a distributor "q"
	M_k^{max}	Maximum financial budget of a manufacturing unit for production of product "k"
	$\tilde{G}$	Fuzzy total amount of production of all types of products

Now we have constructed a Table 20.2 of the following minimum objective functions under given system of constraints of production-distribution planning along with their explanation.

Table 20.2 of Minimize various expenses arising during production-distribution planning under imprecise demands and production quantity by using fuzzy optimization approach

Serial No.	During production	Explanation
1.	$\sum_{k=1}^{K} b_k X_k$	Minimize the total production cost to produce all types of products
2.	$\sum_{k=1}^{K} r_k X_k$	Minimize the total cost of damaged raw material during production
3.	$\sum_{k=1}^{K} l_k X_k$	Minimize total electricity cost for manufacturing of all types of products
4.	$\sum_{k=1}^{K} w_k X_k$	Minimize total worker cost during working hours for production

Serial No.	During production	Explanation
5.	$\sum_{k=1}^{K} h_k X_k$	Minimize total holding cost of inventory of all the manufactured product
6.	$\sum_{p=1}^{P} a_k X_k$	Minimize total holding cost of inventory of all the raw materials
7.	$\sum_{k=1}^{K} u_k X_k$	Minimize total packaging cost of all types of products
8.	$\sum_{k=1}^{K} o_k X_k$	Minimize total worker's overtime cost for production
9.	$\sum_{k=1}^{K} m_k X_k$	Minimize the total worker cost spending for packaging of all types of products
10.	$\sum_{p=1}^{P} C_p A_p$	Minimize total cost of raw materials provided by all the suppliers
	During distribution	
11.	$\sum_{q=1}^{Q} v_q S_q$	Minimize total vehicle maintenance expenses of all the distributors
12.	$\sum_{q=1}^{Q} d_q S_q$	Minimize total packaging cost of all the distributors
13.	$\sum_{q=1}^{Q} c_q S_q$	Minimize total fuel consumption cost during transportation to a market
14.	$\sum_{q=1}^{Q} f_q Y_q$	Minimize total fuel consumption cost during transportation to all distributors
15.	$\sum_{q=1}^{Q} e_q S_q$	Minimize total holding cost of all inventories in all distributors
16.	$\sum_{q=1}^{Q} n_q Y_q$	Minimize total cost allotted to all drivers for transporting to all distributor
17.	$\sum_{q=1}^{Q} t_q S_q$	Minimize total cost allotted to all drivers for transporting product from all distributors to a market
18.	$\sum_{q=1}^{Q} j_q Y_q$	Minimize total cost of damaged product during transportation to all distributors
19.	$\sum_{q=1}^{Q} g_q S_q$	Minimize total cost of damaged product during transportation to a market
	Constraints	
20.	$\sum_{q=1}^{Q} X_k \cong \tilde{G}$	Fuzzy total amount of production of all types of products
21.	$\sum_{p=1}^{P} C_p A_p \leq B^{max}$	Manufacturer restricted the maximum budget for purchasing all raw materials
22.	$\sum_{p=1}^{P} A_p \cong \tilde{R}$	Fuzzy total requirement of all raw materials which is fulfilled by all the suppliers
23.	$\sum_{q=1}^{Q} S_q \cong \tilde{F}$	Fuzzy total demand of the market which is fulfilled by all the distributors together
24.	$\sum_{q=1}^{Q} Y_q \cong \tilde{J}$	Fuzzy total demand of all the distributors which is fulfilled by a manufacturer
25.	$X_k \leq M_k^{max}$	Manufacturer has restricted the maximum budget for production of product "k"
26.	$Y_q \leq D_q^{max}$	Each distributor "q" restricted its maximum financial budget
27.	$X_k \geq 0, \quad Y_q \geq 0$, $S_q \geq 0 \quad A_p \geq 0$ for $k=1,2.....K, \quad q=1,2...Q$, $p=1,2....P$	All decision variables are integers and are greater than or equals to zero

3. Linear membership function for minimum objective functions and fuzzy constraints

The linear membership function for minimum objective function is expressed by equation (28) (Zimmermann (1978)).

$$\mu_{Z_g}(y) = \begin{cases} \frac{Z_g^{**} - Z_g(y)}{D_g(Z)} & \text{if } Z_g^* \leq Z_g(y) \leq Z_g^{**} \\ 1 & \text{if } Z_g(y) \leq Z_g^* \\ 0 & \text{if } Z_g(y) \geq Z_g^{**} \end{cases} \tag{28}$$

The lower and upper bounds of all the objective functions $Z_g(y)$ in its minimization form is represented by Z_g^* and Z_g^{**} for all g=1, 2, 3.........G. On the other hand, the difference between the upper and lower bounds of required objective functions $Z_g(y)$ is represented as $D_g(z)$.

The linear membership function for fuzzy constraint $\tilde{C}$ which is defined by equation (29)

$$\mu_{Z_c}(y) = \begin{cases} \frac{S_q - A_q(y)}{S_q - b_q} & \text{if } b_q \leq A_q(y) \leq S_q \\ 1 & \text{if } A_q(y) \leq b_q \\ 0 & \text{if } S_q \leq A_q(y) \end{cases} \tag{29}$$

where S_q is defined as $b_q + t_q$ and $t_q > 0$ is the level of tolerance, for all q=1, 2........., Q.

Now the given fuzzy multi-objective linear programming model is transformed into a crisp form which is expressed by equations (30–35) (Zimmermann (1978)).

Maximize β   (30)

Subject to,

$$\beta \leq \mu_{Z_g}(y) \qquad \text{for all g=1, 2........., G} \tag{31}$$
$$\beta \leq \mu_{Z_c}(y) \qquad \text{for all q=1, 2.........Q} \tag{32}$$
$$Ay \leq b \qquad \text{for all deterministic nature} \tag{33}$$
$$y \geq 0 \qquad \text{and are integers} \tag{34}$$
$$\beta \in [0,1] \tag{35}$$

Now according to Zimmermann (1978) the optimum upper bound (Z**) and lower bound (Z*) in a proposed fuzzy multi-objective linear programming model is obtained by calculated the maximum and minimum values of same objective function (Z) under the same set of constraints.

4. Numerical example

We have to consider that a manufacturing unit produced two types of products k = 1.2 and considered the total demand of all the raw materials delivered by two suppliers p = 1.2 is 5,000 units and considered the fluctuation in demand of raw materials is from 4,500 units to 5,500 units and also considered the financial budget for purchasing the raw materials is 13,000 rupees. Assuming the total amount of production is approximately about 5,000 units and consider the variation in total amount of production is from 4,000 to 6,000 units. Now in this model we have predicted two distributors q = 1.2 and the total demand of both distributors which is fulfilled by a manufacturer is approximately about 6,000 units which is fluctuated from 5,500 to 6,500 units and the total market demand which is fulfilled by both distributors together is about 7,000 units and assumed the variation in total demand of a market is from 6,500 to 7,500 units.

Now we have constructed the following Table 20.3 by using the notations provided by Table 20.1.

Table 20.3 of Minimize various expenses arising during production-distribution planning under imprecise demands and production quantity by using fuzzy optimization approach

Manufactured product "k"	b_k		r_k	l_k	w_k	h_k	a_k	u_k	o_k	m_k
Product (k=1)	6		2	2	3	5	2	5	3	3
Product (k=2)	20		5	3	2	2	1	3	2	4
Distributor "q"	v_q (one month)	d_q	c_q	f_q	e_q	n_q	t_q	j_q	g_q	
Distributor(q=1)	500	1.5	300	400	4	140	200	100	140	
Distributor(q=2)	300	3	200	500	2	150	320	150	130	

Per unit cost of raw material from suppler "p" (C_p)	Maximum financial budget of a distributor "q" (D_q^{max})	Maximum financial budget for production of product "k" (M_k^{max}).
$C_1 = 2$	$D_1^{max} = 8,000$	$M_1^{max} = 6,000$
$C_2 = 1.5$	$D_2^{max} = 7,000$	$M_2^{max} = 3,000$

Consider the following costs in (Indian Rupees) arises during production-distribution planning.

Now we have constructed a numerical model with the help of the quantitative data which is considered in Table 20.3 and simultaneously minimize all the objective functions defined in Table 20.2 under the given set of constraints.

$Z_1 = 6X_1 + 20X_2$
$Z_2 = 2X_1 + 5X_2$
$Z_3 = 2X_1 + 3X_2$
$Z_4 = 3X_1 + 2X_2$
$Z_5 = 5X_1 + 2X_2$
$Z_6 = 2X_1 + 1X_2$
$Z_7 = 5X_1 + 3X_2$
$Z_8 = 3X_1 + 2X_2$
$Z_9 = 3X_1 + 4X_2$
$Z_{10} = 2A_1 + 1.5A_2$
$Z_{11} = 500S_1 + 300S_2$
$Z_{12} = 1.5S_1 + 3S_2$
$Z_{13} = 300S_1 + 200S_2$
$Z_{14} = 400Y_1 + 500Y_2$
$Z_{15} = 4S_1 + 2S_2$

$Z_{16} = 140Y_1 + 150Y_2$
$Z_{17} = 200S_1 + 320S_2$
$Z_{18} = 100Y_1 + 150Y_2$
$Z_{19} = 140S_1 + 130S_2$

Subject to,
$X_1 + X_2 \cong \overline{5,000}$
$A_1 + A_2 \cong \overline{5,000}$
$S_1 + S_2 \cong \overline{7,000}$
$Y_1 + Y_2 \cong \overline{6,000}$
$2A_1 + 1.5A_2 \leq 13,000$
$\quad X_1 \leq 6,000;$
$\quad X_2 \leq 3,000;$
$\quad Y_1 \leq 8,000;$
$\quad Y_2 \leq 7,000;$

$X_k \geq 0$, $Y_q \geq 0$, $S_q \geq 0$, $A_p \geq 0$, for all k=1, 2; q=1, 2; p=1,2 and are integers.
Now according to Zimmermann in (1978), we have constructed a Table 20.4 in which we have obtained the lower and upper bounds of all the objective functions $(Z_1, Z_2, Z_3, \ldots\ldots\ldots\ldots Z_{19})$ of a given numerical model under the given system of constraints as considered all constraints are deterministic in nature.

Table 20.4 of Minimize various expenses arising during production-distribution planning under imprecise demands and production quantity by using fuzzy optimization approach

Objective function	Upper bound (Z**)	Lower bound (Z*)
Z_1	72,000	30,000
Z_2	19,000	10,000
Z_3	13,000	10,000
Z_4	15,000	12,000
Z_5	25,000	16,000
Z_6	10,000	7,000
Z_7	25,000	19,000
Z_8	15,000	12,000
Z_9	18,000	15,000
Z_{10}	10,000	7500
Z_{11}	35,00,000	21,00,000
Z_{12}	21,000	10,500
Z_{13}	21,00,000	14,00,000
Z_{14}	30,00,000	24,00,000
Z_{15}	28,000	14,000
Z_{16}	9,00,000	8,40,000
Z_{17}	22,40,000	14,00,000
Z_{18}	9,00,000	6,00,000
Z_{19}	9,80,000	9,10,000

Now converted all the traditional minimum objective functions and system of constraints into the fuzzy objective functions and fuzzy constraints by utilizing the linear membership functions defined by equations 28 and 29 and substitute it in equations 30–35.

Maximize β

Subject to,

$$\beta \leq \frac{72000-(6X_1+20X_2)}{42000}$$

$$\beta \leq \frac{19000-(2X_1+5X_2)}{9000}$$

$$\beta \leq \frac{13000-(2X_1+3X_2)}{3000}$$

$$\beta \leq \frac{15000-(3X_1+2X_2)}{3000}$$

$$\beta \leq \frac{25000-(5X_1+2X_2)}{9000}$$

$$\beta \leq \frac{10000-(2X_1+1X_2)}{3000}$$

$$\beta \leq \frac{25000-(5X_1+3X_2)}{6000}$$

$$\beta \leq \frac{15000-(3X_1+2X_2)}{3000}$$

$$\beta \leq \frac{18000-(3X_1+4X_2)}{3000}$$

$$\beta \leq \frac{10000-(2A_1+1.5A_2)}{2500}$$

$$\beta \leq \frac{3500000-(500S_1+300S_2)}{1400000}$$

$$\beta \leq \frac{21000-(1.5S_1+3S_2)}{10500}$$

$$\beta \leq \frac{2100000-(300S_1+200S_2)}{700000}$$

$$\beta \leq \frac{3000000-(400Y_1+500Y_2)}{600000}$$

$$\beta \leq \frac{28000-(4S_1+2S_2)}{14000}$$

$$\beta \leq \frac{900000-(140Y_1+150Y_2)}{60000}$$

$$\beta \leq \frac{2240000-(200S_1+320S_2)}{840000}$$

$$\beta \leq \frac{900000-(100Y_1+150Y_2)}{300000}$$

$$\beta \leq \frac{980000-(140S_1+130S_2)}{70000}$$

$$\beta \leq \frac{6000-(X_1+X_2)}{1000}$$

$$\beta \leq \frac{(X_1+X_2)-4000}{1000}$$

$$\beta \leq \frac{5500-(A_1+A_2)}{500}$$

$$\beta \leq \frac{(A_1+A_2)-4500}{500}$$

$$\beta \leq \frac{7500-(S_1+S_2)}{500}$$

$$\beta \leq \frac{(S_1+S_2)-6500}{500}$$

$$\beta \leq \frac{(Y_1+Y_2)-5500}{500}$$

$$\beta \leq \frac{6500-(Y_1+Y_2)}{500}$$

$2A_1 + 1.5A_2 \leq 13000$
$X_1 \leq 6000;$
$X_2 \leq 3000;$
$Y_1 \leq 8000;$
$Y_2 \leq 7000;$

$X_k \geq 0, Y_q \geq 0, S_q \geq 0, A_q \geq 0$, for all k=1, 2; q=1, 2; p=1,2 and are integers and also satisfied the condition $0 \leq \beta \leq 1$.

Now with the help of linear programming-based software LINGO (Ver 20.0) we have solved the above numerical model and the obtained solution which is defined as follows:

$\beta = 0.5481429, X_1 = 3656, X_2 = 893; A_1 = 2934, A_2 = 1841; S_1 = 3387, S_2 = 3388;$
$Y_1 = 2614, Y_2 = 3161$

So, the values of all the objective functions are defined as follows:

$Z_1 = 39,796,\quad Z_2 = 11,777,\quad Z_3 = 9,991,\quad Z_4 = 12,754,\quad Z_5 = 20,066,\quad Z_6 = 8,205,$
$Z_7 = 20,959,\quad Z_8 = 12,754,\quad Z_9 = 14,540,\quad Z_{10} = 8,629.5,\quad Z_{11} = 2,709,900,$
$Z_{12} = 15,244.5,\quad Z_{13} = 1,693,700,\quad Z_{14} = 2,626,100,\quad Z_{15} = 20,324,\quad Z_{16} = 474,150,$
$Z_{17} = 1,761,560, Z_{18} = 735,550, Z_{19} = 914,620.$

5. Conclusion

In this research study we have constructed fuzzy multi-objective linear programming model through which we have minimized different types of expenses arises during production-distribution planning such as expenses of raw materials, expenses related to packaging, inventory holding related expenses, fuel consumption related expenses, workforce related expenses and so on under different imprecise demands which involve total demand of raw materials, total market demand, total demand of all the distributors as well as total production quantity which is also imprecise in nature. Finally, we have presented the practical applicability and validity of this research study by taking a numerical example and solve it by using a linear programming-based software LINGO (Ver. 20.0).

References

1. Ghanbari, R., Ghorbani-Moghadam, K., Mahdavi-Amiri, N., and De Baets, B. (2020). Fuzzy linear programming problems: Models and solutions. *Soft Comput.*, 24(13), 10043–10073.
2. Goodarzian, F. and Hosseini-Nasab, H. (2021). Applying a fuzzy multi-objective model for a production–distribution network design problem by using a novel self-adoptive evolutionary algorithm. *Int. J. Sys. Sci. Oper. Log.*, 8(1), 1–22.
3. Ghanbarzadeh-Shams, M., Yaghin, R. G., and Sadeghi, A. H. (2022). A hybrid fuzzy multi-objective model for carpet production planning with reverse logistics under uncertainty. *Socio-Economic Plan. Sci.*, 83, 101344.
4. Seif Barghi, M. and Shirin Bayan, P. (2023). Fuzzy multi-objective production distribution planning by considering CO2 emission cost and solving by a novel fuzzy multi-choice goal programming. *J. Indus. Manag. Perspec.*, 13, 173–198.
5. Zimmermann, H. J. (1978). Fuzzy programming and linear programming with several objective functions. *Fuzzy Sets Sys.*, 1(1), 45–55.

21 Queuing theory and communication networks

Satish Chandra Tiwari[1,a] and Sanjay Chaudhary[2,b]

[1]Associate Professor, Retd. Department of Mathematics, S.V. College, Aligarh

[2]Prof. & Head, Department of Mathematics, Institute of Basic Sciences, Dr. B. R. Ambedkar University, Khandhari, Agra, Uttar Pradesh, India

Abstract

The study made in this paper gives an insight into the inter-relationship of queuing and communication systems/networks, which strongly recommend that queuing theory can efficiently be employed for communication networks. The study has been justified by the survey of several previous studies made in this direction and by designing a computer/communication network.

Keywords: Queuing theory, communication systems, networks

Introduction

A "Queue" or a waiting line, is a group of arriving items (units or customers) that wait to be served at the service station (service facility) which provides the service they seek.

Queuing theory is mostly a methodological approach to study congestion phenomenon. It attempts to formulate interpret and seeks optimal solution to the congestion problems subject to client and server constraints. The theory is being proved to be very successful to provide queuing models to understand different complex manufacturing and service systems including communication networks.

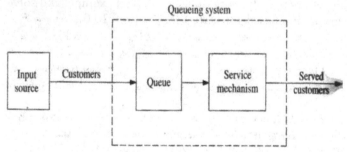

A basic queuing process

Queuing process is a process in which customers reach the service station for the service they seek and join the queue(s). A queue and corresponding customers are selected for a service as per pre-defined queue discipline/service mechanism.

[a]sctsvca@gmail.com, [b]scmibs@hotmail.com

Queue discipline refers to the selection criteria/pattern which decides to select particular queue/customers for the service they require, it may include **first-come-first-served (FCFS)**, **last-come-first-served (LCFS)**, service in random order (SIRO) and some pre-decided priority rules.

Service mechanism refers to the pre-defined rules and protocols, characterized by the service provider for the network.

Queuing systems are described by the arrival and service patterns/distributions, the number of serving channels, pre-decided service rules capacity of the system and the capability of the calling source.

Queuing networks are the systems of multiple queues connected via different channels. When the service of a customer is completed on a node/queue, it changes as per its requirement and available facility and finally leaves the system. The queuing networks are categorized as **open, closed and mixed** queuing networks. The important queuing networks to mention but a few are :

Jackson Networks,
Gordon Networks,
Kelly's Networks,
BCMP (Baskett, Chandy, Muntz and Palacius) Networks,
Fork-join Networks.

Designing a queuing network primarily involves the study of different queuing systems constituting networks, their arrival patterns and then to decide/specify the service patterns, topologies and priority rules accordingly under the limitations of control.

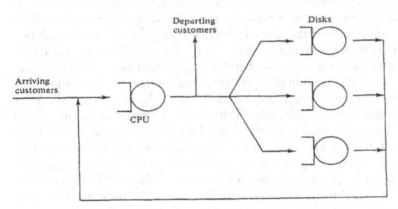

A queuing/communication network

Communication network is identified as a network of multiple queues and each queue as a collection of data packets waiting for communication/transmission by a network of terminal nodes, connected together with the **links**. The nodes share the data through appropriate links using either of the processes – circuit switching, packet switching or message switching.

The important communication networks include but are not limited to:

- Computer networks
- Internet
- The telephone network
- The global telex network
- The aeronautical **ACARS** network.

History / literature review

Allen (1980) studied queuing networks and computer systems deeply and explained their inter-relationship and showed efficiently through a number of examples how we can use queuing theory to create analytical models of computer systems. These models are easy to use and efficient to provide better results with a little effort. He remarked "Benchmark is older, simulation is more popular". But for computer systems performance, queuing models may provide the most cost-effective models/techniques. Modiano et al. (1996) discussed discrete-time queues and found the expression for the average delay. The service time distribution in this study has been taken as constant and identical for each queue. The result is applicable for packet management where data packets are of fixed duration. Filipowicz and Kwiecien (2008) studied different models and their applications to various systems e.g., computers, communications, transportation and industries, etc. The models formulate successfully a number of queuing systems and networks. They studied the models for both Markovian; Poisson arrival, exponential service time distributions and non-Markovian with Cox and Weibull service time distributions. In this study, they also discussed different queuing networks e.g., Jackson – network, Gordon – Newell networks, Kelly networks, BCMP Network and Fork–join systems. They analyzed the performance measures, utilization, throughput and response times for different systems/models. Arnaud et al. (2009) studied the smart intelligent transport systems for enhancing sensing capabilities, safety and efficiency of the network. In this regard, they surveyed a number of projects, related technologies, approaches and the solutions. Neeley (2010) discussed in the text "stochastic network optimization with applications to communication and queuing systems" dynamic networks, their performance control and optimization. In the study, the throughput and related measures have been optimized using the drift-plus penalty method. Torkey et al. (2012) proposed a fast recovery algorithm/mechanism to enhance the congestion control performance of TCP (transmission control protocol). The mechanism has been developed adapting the congestion window of the TCP in the network. They evaluated the algorithm by comparing the results obtained by simulation and found that this mechanism provides better results for both packet delay and throughput. Dike et al. (2016) used queuing theory to solve the congestion control problem of data communication networks. They preserved a queuing model which enables the planning engineer to plan an effective and efficient network. Maguluri et al. (2018) studied internet and data center networks in the queue – theoretic terms. They discussed the saturation situation of the posts (ports), used the drift technique with Max – Weight scheduling algorithm, found queue length and other optimal measures. Bandi et al. (2018) studied the transient behavior of the multiserver and feed-forward queuing networks. The analytical approach they used is based on the concept, of a polyhedral set characterized by the variability parameters of the model. The results obtained provide a deep qualitative insight into the system times and essential measures of the networks. Baron et al. (2018) studied a state-dependent single server (M/G/1) Queuing system. In this study, they introduced a probabilistic Markov chain decomposition approach for performance measures. They also discussed the applicability of the approach to other non-Markovian systems. This study is useful where the customers do not present physically but have remote access e.g., a call center or a communication node. Atar et al. (2019) proposed a load-balancing policy for heavy traffic. This policy may be considered to lay b/w the policies JSQ, join the shortest queue and JLW, join the least workload. Under this policy, the arrivals into the buffers are replicated into the tasks and are directed to join the shortest queue. This policy abbreviated as RSQ, replicates to the shortest queues. They showed that as the number of tasks performing queues increases server, workloads become more balanced and the waiting tail becomes smaller. Roy et al. (2019) studied a communication network in queue theoretic terms. In the study they considered the sources providing updates to the destination through the simple queues. Kesidis et al. (2020) studied the measure, age – of – information (AOI) considering as a stochastic process in the stationery environment and characterized the distribution using the palm inversion formula. They used the system with small or no buffer and renewal theory assumptions to find the explicit solution for the Laplace transform. Cruise et al. (2020) studied a queuing network where the arrival pattern is Poisson and the newly arriving customer is directed to join the queue with the shortest length. In this study, they showed that the system achieves stability under the defined specific routing policy. This policy is like a policy to join the queue of shortest workload (JSW). Glynn (2022) While throwing light on modeling in queuing theory after having an insight, they discussed three types of models; descriptive, prescriptive and predictive. Descriptive models are intended to have quantitative insight into a queuing phenomenon, while prescriptive modeling focuses on optimal design and control while in the third type of modeling i.e., in predictive focus is given on accuracy to predict real-life phenomenon. He pointed out that most of the historical queuing models have focused on descriptive and prescriptive models but with the development of tools and technologies there is a need to focus on predictive modeling which can transform real-life problems to study and predict effectively. Mittal and Sharma (2022) developed the queuing model to study the communication flow. For the study, they also reviewed the fuzzy set theory of the concern. In this study, the arrival and service patterns together with covariance are fuzzy in nature. They used the defuzzification process to explore system characteristics. For the validation of results, they used Numerical and graphical methods. In his study, Arshi Naim (2023) made an attempt for the retailers/small storekeepers how they can run their stores effectively by maintaining social distancing under the existing physical environment, which has been a burning issue during Covid-19 pandemic. The study is based on electronic queuing systems and is helpful to achieve the goal. George et al. (2023) in their study considered AOI as an important performance measure to provide freshness of the information. They studied the system with the arrival process as Poisson and the service time distribution as general independent and used Markov-renewal theory to derive results. The study suggests the solution to the optimality problems regarding previous scheduling policies and also discusses the management of AOI.

Designing a communication network: As in a computer/communication network, **a Queue** can absolutely be defined as a collection of data packets waiting to be transmitted by a network device using a predefined service mechanism which can be identified as a queuing network. Hence queuing network models can efficiently be employed for communication networks also. Though the communication has been an inevitable part of our life but now with the advancement of science and technology the human society is going on dependent more and more on communication systems. They are affecting our lifestyle to a great extent and it is true that **"Smarter the communication system, Smarter the lifestyle"**. While designing a queuing network/communication network, we try to optimize the **performance measures**. Some performance measures are said to be **user oriented** and some are said to be **system oriented**. Therefore our aim must be to meet a golden mean between these two. **Queuing theory** is sufficiently developed to deal with a number of communication networks. Communication networks are the LANS or collection of LANS consisting of physical networks – interconnected computers and channels, messages flow (having origin, destination etc.) and operating rules. The LANS are the collection of nodes, routers, communication channels, etc., connected via gateways and backbone. A conceptual computer/communication network has been designed and illustrated by a pictorial diagram/flow chart, given below:

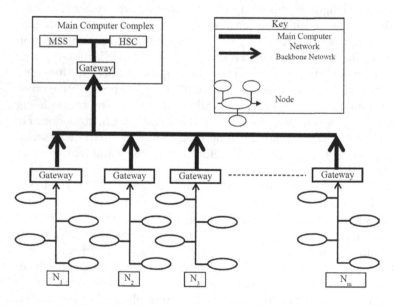

A conceptual communication network

The smart communication network needs intelligent planning and modeling which involves understanding/studying the individual LAN(s) together with various components constituting the network and the network as a whole with the goal to achieve optimal performance measures establishing a happy medium (trade-off) between the user and the system. Some of the important components/parameters to study include but are not limited to:

- Data generation/transmission/service rates and patterns over each path
- Load and capacity of the network
- Effect of parameter(s) changing on performance measures.

Applications

The queuing theory first evolved early in the 20th century with the applications in telephony and communication systems by the pioneer A. K. Erlang (1909), a Danish telephone engineer. Since its beginning, it is going on spreading over quite a few fields of study in a wide variety of contexts: modeling for information, manufacturing and service systems, man and machine performance and for chemotherapy units to quote but a few.

Conclusion

With the population explosion and advancement of science and technology computer systems and communication has become an inevitable part of our life and requires a smart predictive type of modeling. The computer communication network is a queuing network – a collection of customers (users/data packets), the service station which represents system resources, hence queuing network models can efficiently be employed for computer communication models also to optimize performance measures - throughput, utilization and response times, etc.

References

1. Allen, A. O. (1980). Queueing models of computer systems. *Mag. Comp.*, 13(4), 13–24.
2. Arnaud, et al. (2009). Vehicular communication systems: Enabling technologies. *Appl. Fut. Outlook Intel. Trans. IEEE Comm. Mag.*, 47(11), 84–95.
3. Arshi, N. (2023). Business to business through application of queueing theory. *Web Res. Int. Sci. Res. J.*, 1(1), 1–12.
4. Atar, R. et al. (2019). Replicate to the shortest queues. *Que. Sys.*, 92, 1–23.
5. Bandi, et al. (2018) . Robust transient analysis of multi-server queueing systems and feed–Forward networks. *Que. Sys.* 89, 351–413.
6. Baron, et al. (2018). The state dependent M/G/1 queue with orbit. *Que. Sys.*, 90, 89–123.
7. Cruise, J. et al. (2020). Stability of JSQ in queues with general server – Job class compatibilities. *Que. Sys.*, 95(1–2), 1–7.
8. Dike, D. O. et al. (2016). Improving congestion control in data communication network using queueing theory model. *IOSR J. Elec. Electron. Engg. (JEEE).*, 11(2), 49–53.
9. Filipowicz, B. and Kwiecien, J. (2008). Queueing systems and networks. *Models Appl. Bull. Polish Acad. Sci.*, 56(4), 379–390.
10. George, K. et al. (2023). Age of information using Markov renewal methods. *Que. Sys.* 103, 95–130.

11. Glynn, P. W. (2022). Queueing theory: Past, present and future. *Que. Sys.*, 100(3), 169–171. http://doi.org/10.1007/s11134-022-09785-4.

12. Kesidis, G. et al. (2020). The distribution of age of information performance measure for message processing system. *Que. Sys.*, 19.

13. Maguluri, S. T. et al. (2018). Optimal heavy traffic queue length scaling in an incompletely saturated switch. *Que. Sys.*, 88, 279–309.

14. Mittal, H. and Sharma, N. (2022). Modeling of communication network with queueing theory under fuzzy environment. *Math. Stat. Engg. Appl.*, 71(2), 122–137.

15. Modiano, E. et al. (1996) . A simple analysis of average queueing delay in tree networks. *IEEE Trans. Inform. Theory*, 42(2), 660–664.

16. Neely, M. J. (2010). Stochastic network optimization with applications to communication and queueing systems. *Syn. Lect. Comm. Netw.*, 3(1), 1–211.

17. Roy, D. et al. (2019). The age of information: Real-time status updating by multiple sources. *IEEE Trans. Inform. Theory*, 65(3), 1807–1827.

18. Torkey, H. et al. (2012). Modified fast recovery algorithm for performance enhancement of TCP-NEW RENO. *Int. J. Comp. Appl.*, 40(12), 30–35.

22 Dynamics of tumor growth with delayed differential equations: Impact of time delays on disease progression and treatment

Diksha Gautam[a], Sanjeev Kumar and Deepak Kumar

Department of Mathematics, Dr. Bhimrao Ambedkar University, Khandari Campus, Agra 282002, Uttar Pradesh, India

Department of Mathematics, Manav Rachna International University, Faridabad, India

Abstract

Cancer is a complex disease that requires an in-depth understanding of its growth dynamics and effective treatment strategies. Mathematical models provide a useful tool for analyzing tumor growth and designing optimal control strategies. In this study, we investigate tumor growth dynamics using delayed differential equations and optimal control theory. The model incorporates tumor growth dynamics, immune response, and the effect of treatment. We aim to identify effective treatment strategies by exploring different control strategies, including chemotherapy, immunotherapy, and a combination of the two. Our findings demonstrate that the most effective approach is optimal control strategy depends on the tumor development stage and the tumor growth delay. We find that delaying treatment may lead to better outcomes in some cases, while early treatment may be necessary in other cases. Our study highlights the importance of considering the delay in tumor growth when designing treatment strategies and provides insights into optimal control strategies for managing cancer. The findings of this study could potentially inform clinical practice and improve the management of cancer patients.

Keywords: Time delay, DDE, tumor, mathematical modeling cytotoxic CD8+ T cell, diffusion, Matlab

Introduction

Cancer constitutes a multifaceted ailment characterized by the unrestrained proliferation and dissemination of aberrant cellular entities. Understanding the dynamics of tumor growth is a critical step in developing effective treatments and improving patient outcomes. Mathematical modeling has been employed to examine the expansion and advancement of tumors. Delayed differential equations (DDEs) are particularly useful in capturing the time lapse between the initiation of a mutation and the manifestation of a tumor. Optimal control theory (OCT) is another mathematical tool that has been increasingly used in cancer research to design optimal treatment strategies. The goal of OCT is to find the best treatment regimen that maximizes the therapeutic effect while minimizing

[a]gautamdiksha333@gmail.com

side effects and toxicity. In this research paper, we focus on modeling tumor growth using DDEs and OCT. DDEs allow us to incorporate time delays into the model, which can capture the effects of biological processes such as cell division and migration. OCT enables us to design treatment plans that minimize tumor growth while minimizing side effects. We begin by formulating a set of DDEs that describe the dynamics of tumor growth, considering the interplay among tumor cells, the immune system, and the microenvironment. Then we use OCT to design treatment strategies that optimize the trade-off between tumor size and treatment-related toxicity. We consider both chemotherapy and immunotherapy as treatment options and investigate how the optimal treatment strategy depends on the tumor growth rate, the patient's immune status, and the toxicity of the treatment. The studies mentioned here are various research papers that have been published on the use of mathematical modeling in understanding tumor growth and designing effective treatment strategies. Dixit and Kumar's team (2014), analyzed the mathematical model to study the solid tumor during the angiogenesis phase accompanied by an immune response. In a subsequent study, Dixit and Kumar team (2015) proposed a mathematical model that incorporated biological stoichiometry to investigate the effects of chemotherapy on tumor growth. Bodnar and team (2007) identified three categories of basic DDEs depicting tumor proliferation. Barbarossa (2012) proposed delay equations that simulated the impact of phase-specific medications and immunotherapy on actively dividing tumor cells. Rihan and colleagues (2016) proposed tumor-immune system dynamics incorporating fractional order. Ghosh and team (2017) analyzed the manner in which tumor growth is affected by postponed interactions between cancer cells and the microenvironment. Rihan and colleagues (2019) investigated the dynamics of a delay differential model with fractional order in relation to the tumor-immune system. Ozkose and colleagues (2022) determined fractional modeling of the interplay between the immune system and tumors, specifically in the context of lung cancer, using authentic data. Rihan and colleagues (2022) analyzed the behavior of a time-delay differential model concerning interactions between tumors and the immune system, incorporating stochastic noise. Overall, the use of mathematical modeling in cancer research provides a powerful tool for gaining a deeper understanding of the intricate dynamics involved in the growth of tumors. and for designing optimal treatment strategies. As ongoing research continues to refine and expand these models, they have the potential to lead to significant advances in cancer treatment and patient outcomes.

Models and formulation

Mathematical models are powerful tools used to describe complex systems and predict their behavior. One such model is the one developed by Ira et al. (2020), which describes the dynamics of tumor growth and the impacts of diverse treatments, including chemotherapy and immunotherapy. The model includes four types of cells: tumor cells, natural killer cells, dendritic cells, and activated immune system cells. By incorporating a systematic control function, which represents the application of natural killer cells combined with CD8+ T cell treatment, the researchers investigated the effect of these treatments on the system parameters. The results showed that immunotherapy significantly impacted the stability of tumor cells, making CD8+T cells more stable and the treatment more effective (Figure 22.1).

$$D_t(T) = \lambda T(1 - \mu T) - (\omega N + \rho D + \gamma L)T - \mathcal{K}_T \Lambda T$$

$$D_t(N) = \upsilon + \mathcal{M}_{mt} - (\omega T - \alpha D)N - \mathcal{K}_N \Lambda N - \mathcal{E}N \tag{1}$$

$$D_t(D) = \beta - (\eta L + \alpha N - \sigma T)D - \mathcal{K}_D \Lambda D - jD$$

$$D_t(L) = \rho DT - \gamma LT - \chi NL^2 + \omega NT + \mathcal{P}_{LI} - \mathcal{K}_L \Lambda L - gL + \upsilon_t$$

With Initial conditions are $T \geq 0, N \geq 0, D \geq 0, L \geq 0, M \geq 0, I \geq 0$

Time delay models are crucial in tumor mathematical modeling as they account for the inherent time delays in the recruitment of immune cells and the interactions between different cell types. These delays have a substantial influence on the dynamics of tumor growth and the efficiency of treatments. Delayed differential equations (DDEs) are commonly used to model these delays, and have been demonstrated to produce more precise predictions than traditional differential equations that do not account for time delays. To incorporate more realistic biological assumptions, we have enhanced the original model (1) by integrating delay differential equations and revising the Michaelis-Menten form. These modifications are grounded in the latest biological research and are expected to yield a more accurate representation of the system being studied. Here are the DDEs corresponding to the system of equations (1), with delays denoted by τ:

$$\frac{dT}{dt} = \lambda T(t - \tau)(1 - \mu T(t - \tau)) - (\omega N(t) + \rho D(t) + \gamma L(t))T(t - \tau) - \mathcal{K}_T \Lambda T(t - \tau)$$

$$\frac{dN}{dt} = \upsilon + \mathcal{M}_{mt}(t - \tau) - (\omega T(t) - \alpha D(t))N(t - \tau) - \mathcal{K}_N \Lambda N(t - \tau) - \mathcal{E}N(t - \tau)$$

$$\frac{dD}{dt} = \beta - (\eta L(t) + \alpha N(t) - \sigma T(t))D(t - \tau) - \mathcal{K}_D \Lambda D(t - \tau) - jD(t - \tau) \tag{2}$$

$$\frac{dL}{dt} = \rho D(t)T(t) - \gamma L(t - \tau)T(t) - \chi N(t)L^2(t - \tau) + \omega N(t)T(t) + \mathcal{P}_{LI}(t - \tau) - \mathcal{K}_L \Lambda L(t - \tau) - gL(t - \tau) + \upsilon_t$$

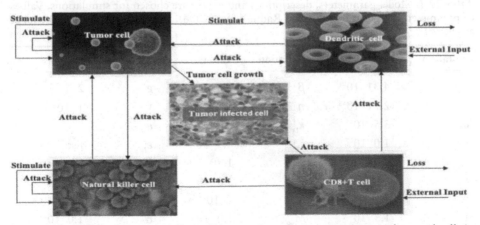

Figure 22.1 Structure to define cell interaction of tumor, immune, and normal cells in the presence of chemotherapy drug

For $\tau = 0$, the model is reduced to an ODE model, which is a simpler version of the original model with delay. The initial data of the ODE model still satisfies the non-negativity constraints on the four variables: $D_0 \geq 0$, $T_0 \geq 0$, $N_0 \geq 0$, *and* $L_0 \geq 0$.

$$T(\emptyset) = \delta(\emptyset), N(\emptyset) = \delta(\emptyset), \qquad\qquad L(\emptyset) = \delta(\emptyset), \qquad D(\emptyset) = \delta(\emptyset),$$

$$T(\emptyset) = \delta_1(\emptyset) \text{ for } \emptyset \in [-\tau, 0] \qquad\qquad L(\emptyset) = = \delta_3(t) \text{ for } \emptyset \in [-\tau, 0]$$

$$N(\emptyset)) = \delta_2(\emptyset)) \text{ for } \emptyset \in [-\tau, 0] \qquad\qquad D(0) = D_0 \tag{3}$$

$$\delta_1(\emptyset) \geq 0, \quad \delta_2(\emptyset) \geq 0, \quad \delta_3(\emptyset) \geq 0, \quad \delta_4(\emptyset) \geq 0 \qquad \delta_1(0) \geq 0, \delta_2(0) \geq 0, \delta_3(0) \geq 0, \delta_4(0) \geq 0$$

The system of equations (2) has a time delay or time lag term denoted by τ. This means that the value of the variable at the current time is dependent on its past value at time $t–\tau$. The initial conditions for the system ensure that all variables are non-negative, which is biologically meaningful. The time delay accounts for the fact that the effect of a certain action may not be felt immediately, but rather after some time has passed (Table 22.1).

Result and discussion

The proposed enhanced tumor growth model, which incorporates delay differential equations and revised Michaelis-Menten form, provides a more realistic and accurate representation of the system under study. The DDEs in the model take into account the inherent time lags in the recruitment of immune cells and interactions between different cell types, which significantly impact of the dynamics of tumor growth and the effectiveness of treatments (Figures 22.2–22.5).

The simulation outcomes also carry significant implications concerning the formulation of efficacious cancer therapies. The enhanced model provides a more

Table 22.1 Model parameters, descriptions and values are chosen for simulations. Values of parameters are given by Pillis and Radunskaya (2006), Trisilowati et al., (2013), Ira et al., (2020).

Notation	Value	Notation	Value	Notation	Value
λ	4.31 10^{-1}	β	4.8 10^2	g	2 10^{-2}
μ	2.71 10^{-8}	η	1 10^{-8}	v_I	1 10^4
ω	3.5 10^{-6}	κ_D	6 10^{-4}	ε	Est.
ρ	1.0 10^{-7}	j	2.4 10^{-2}	ε_I	Est.
γ	3.42 10^{-10}	x	1.80 10^{-8}	lM	Est.
κ_T	9 10^{-4}	P_{LI}	3.0 10^3	cI	Est.
Λ	9 10^{-1}	K_L	6 10^{-4}	ε	4.12 10^{-2}
υ	1.3 10^4	M_{mt}	1.0 10^4	σ	1.0 10^{-4}
T	100 cells	α	4 10^{-6}	κ_N	6 10^{-4}

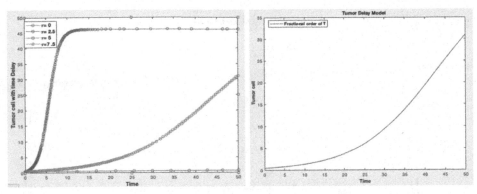

Figure 22.2 Numerical simulations of tumor cell for time delay (τ) with initial condition (0, 2.5, 5, 7.5) to satisfies the non-negativity of the system of equation (2)

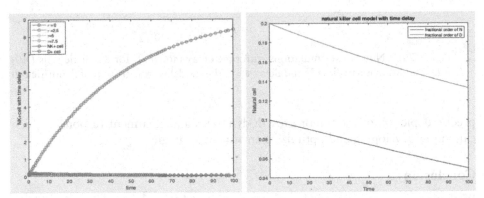

Figure 22.3 Numerical simulations for NK+ cells delay (τ) with initial condition (0, 2.5, 5, 7.5) satisfies the non-negativity of the system of equation (2)

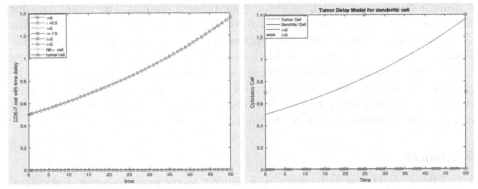

Figure 22.4 Numerical simulations of the time delay model (2) for CD8+T cells delay τ with initial condition (0, 2.5, 5, 7.5) satisfies the non-negativity of the system of equation (2)

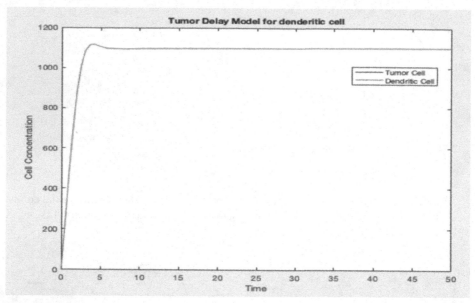

Figure 22.5 Numerical simulations with time delay model (2) for dendritic cells $D(0)$ = D_0 condition is satisfies (3) and delay $\tau = 0$, the model is stop growth of dendritic cell

precise depiction of the tumor growth dynamics and treatment response, which can aid in developing and optimizing treatment strategies.

Conclusion

We have presented a modified mathematical model for tumor growth that incorporates time delay using delayed differential equations (DDEs). The model considers the interactions between different cell types, recruitment of immune cells, and the effects of treatments on tumor growth. The modifications made to the model are grounded in the latest biological research and are expected to yield a more accurate representation of the system being studied. The initial data for the model satisfies the non-negativity constraints on the four variables. The time delay accounts for the fact that the effect of a certain action may not be felt immediately but rather after some time has passed. The model can be further improved by incorporating more realistic biological assumptions and data. Overall, the proposed model provides a useful tool for studying tumor growth and the effectiveness of various treatments. This could eventually contribute to formulating enhanced therapeutic approaches for individuals with cancer.

References

1.   Pillis, L. G. D. and Radunskaya, A. E. (2006). Mixed immunotherapy and chemotherapy of tumors: modelling, application and biological interpretations. *J. Theoret. Biol.*, 238(4), 841–862.

2. *Bodnar, M. and* Forys, U. (2007). Three types of simple DDE's describing tumor growth. *J. Biol. Sys.,*15(4), 453–471.

3. Barbarossa, M. V., Kuttler, C., and Zinsl, J. (2012). Delay equations modeling the effects of phase-specific drugs and immunotherapy on proliferating tumor cells. *Math. Biosci. Engg.*, 9(2), 241–257.

4. Trisilowati, Scott, W. M., and Dann, G. M. (2013). An optimal control model of dendritic cell treatment of a growing tumor. *Anziam*, 54(2012), 664–680.

5. Dixit, S., Kumar, D., Kumar, S., and Johri, R. (2014). Mathematical model solid tumor at the stage of angiogenesis with immune response. *Int. J. Innov. Sci. Engg. Technol.*, 1(7), 174–180.

6. Dixit, D. S., Kumar, D., Kumar, S., and Johri, R. (2015). Mathematical modelling for chemotherapy of tumor growth with an aspect of biological stoichiometry. *Global J. Pure Appl. Math.*, 11(4), 2581–2587.

7. Rihan, F. A., Hashish, A., Al-Maskari, F., Sheek-Hussein, M., Ahmed, E., Riaz, M. B., and Yafia, R. (2016). Dynamics of tumor-immune system with fractional-order. *J. Tumor Res.*, 2(1), 01–06.

8. Ghosh, D., Khajanchi, S., Mangiarotti, S., Denis, F., Dana, S. K., and Letellier, C. (2017). How tumor growth can be influenced by delayed interactions between cancer cells and the microenvironment? *BioSys.*, 158, 17–30.

9. Rihan, F. A. and Velmurugan, G. (2019). Dynamics of fractional-order delay differential model for tumor-immune system. *Chaos Solitons Fract.*, 132(3), 01–14.

10. Ira, J. I., Islam, M. S., Misra, J. C., and Kamrujjaman, M. (2020). Mathematical modelling of the dynamics of tumor growth and its optimal control. *Int. J. Ground Sed. Water*, 11, 659–679.

11. Ozkose, F., Yılmaz, S., Yavuz, M., Ozturk, I., Senel, M. T., Bagcı, B. S., Dogan, M., and Onal, O. (2022). A fractional modeling of tumor–immune system interaction related to lung cancer with real data. *Eur. Phy. J. Plus,* 137(1), 40.

12. Rihan, F. A., Alsakaji, H. J., Kundu, S., and Mohamed, O. (2022). Dynamics of a time-delay differential model for tumor-immune interactions with random noise. *Alexandria Engg. J.*, 61(12), 11913–11923.

23 Forecasting milk and egg production in India: A comparative study of different deterministic growth models

Bharti[a] and Meenakshi Srivastava[b]

Department of Statistics, Institute of Social Sciences, Dr. Bhimrao Ambedkar University, Agra-282004, Uttar Pradesh, India

Abstract

To a large extent, India's national economy depends on the livestock industry. India has a tremendous supply of livestock. The sustenance of approximately 20.5 million people depends on livestock. The purpose of the study is to forecast milk and egg production in India using linear, exponential, quadratic, compound, and cubic deterministic growth models and also using the most recent model selection criteria to find the best model which could most appropriately portray the growth trend of production of milk and egg in India. For this purpose, the time series data (1981–1982 to 2018–2019) on milk and egg production from "Basic Animal Husbandry and Fisheries Statistics–2019, Government of India", was taken into account. The statistical analysis shows that the cubic model works best for predicting both milk and egg production. The selected model is applied for forecasting milk and egg production over the next 10 years with a 95% confidence interval. The policy implication of the present research is that it will assist the planners in taking appropriate decisions for improving management practices as per demand and supply of milk and egg production.

Keywords: Forecasting, livestock production, deterministic growth model, model selection criteria

1. Introduction

India's national economy would suffer greatly without the contributions of the livestock industry. The forecasting of milk and egg production has recently gained a lot of attention for the purposes of planning and policymaking. According to the 20th livestock census, the livelihood of almost 20.5 million people depends on livestock. Compared to an average of 14% for all rural households for two-thirds of rural communities, livestock is the main source of income. Additionally, it employs around 8.8% of India's population. India has vast livestock resources. The livestock sector contributes 4.11% of the GDP and 25.6% of the total agriculture GDP. For healthy physical and mental development of human beings high, quality animal protein which is present in eggs and milk, is an essential requirement. Research studies indicate that due to significant benefits, milk and egg are one of the most widely consumed sources of animal protein in India. The per capita availability of milk is around 375 g / day; eggs are 74 / annum during 2017–2018.

As per the survey report published by the Gujarat Cooperative Milk Marketing Federation (Times of India March 15, 2021), the production of milk and eggs

---

[a]bhartichauhan1182@gmail.com,[b]msrivastava_iss@hotmail.com

in the state of Gujarat in 2018–2019 has been the highest in the past 15 years, according to the latest socio-economic data. Gujarat registered egg production at 18,544 lacks and milk production at 144.9 lakh tones during this year. Uttar Pradesh is the highest milk-producing state and contributes around 18% of the total milk produced; Andhra Pradesh occupies the top position in egg production and contributes over 19% of total production.

The best forecast of the production pattern of milk and egg can be made using an appropriate time series model, which will well describe the observed data. Time series models of two types are generally in vogue for forecast purposes, namely (i) stochastic time series models and (ii) deterministic time series models. Stochastic time series models present data and predict outcomes incorporating features of randomness or unpredictability. The deterministic time series models are inexpensive, provide quick estimates, and, above all, are easy to understand in many situations. Some of the commonly used deterministic time series models are cubic, logarithmic, quadratic, linear, S-shaped, exponential, compound and inverse power. According to the definition used by Pindyck and Rubinfeld (1998), these models are termed deterministic because they don't discuss where and how the series' inherent randomness comes from. The growth rate of the time series can be estimated with the help of these models. Finding the most appropriate growth model for the time series is a prerequisite for conducting growth analysis. Using the most up-to-date model selection parameters (R^2, RMSE, AIC, BIC, MAE, and MAPE), this article seeks to determine the most effective model for milk and egg production in India. Also, the best-fitted model is used to characterize the growth scenario and predict milk and egg production in India.

2. Materials and methods

The data used for the present study is secondary data. It is collected from Basic Animal Husbandry and Fisheries Statistics 2019, Government of India from 1981–1982 to 2018–2019. Data is related to milk and egg production availability in India collected for the above-mentioned period. Statistical analysis has been performed using "SPSS" Statistical Software. Growth models are expressed in form of mathematical equations. They are used to elucidate the pattern of growth through integrated variable which varies with time. The behavior of the integrated variable can describe systematic variation or a particular trend. The trend can be deterministic or stochastic. If the trends are completely predictable, they are termed deterministic trends.

It is to be recalled that these models are called deterministic because they do not provide any information regarding the nature of randomness in the series, and also, in them, no reference is made to the source. These types of deterministic time series models are often used for forecasting purposes. The applicability of these types of models depends on the type of growth that occurs in time series data in a specific area of a specific problem. They are also termed growth models.

The deterministic trend can be defined in terms of any function of time. For example

GT= a+bt   (linear trend)

$GT= \sum_{k=0}^{i} \partial_k \, t^k$ (polynomial time trend)

Linear, exponential, and compound models all have stable rates of growth. With a linear model, the annualized rate of development remains the same throughout the entire observation period. The growth rates are constant over time in linear, exponential, and compound models. However, in the expansion of other types of models, growth is time-dependent. The major difference between compound and exponential models is the choice and understanding of β coefficient values (Granger and Newbold (1986); Gujarati (2003)). The results of these models' forecasts are often merged with those of other models to produce more accurate predictions (Table 23.1).

Table 23.1 Growth models – algebraic structures

Model	Algebraic structure	Interpretation of notations
Linear	$Y_t = a + bt + \epsilon$	Y_t: time series considered t denotes
Quadratic	$Y_t = a + bt + ct^2 + \epsilon$	time which takes integer values starting from 1
Cubic	$Y_t = a + bt + ct^2 + dt^3 + \epsilon$	ϵ represents residual regression a, b, and c are the coefficients of the model
Compound	$Y_t = ab^c e^\epsilon$	
Exponential	$Y_t = ae^{bt\epsilon}$	

Model selection criteria: Model selection is a crucial step in any statistical analysis. Most methods for picking one econometric model over another involve the following steps: estimate each econometric model using a technique for solving an optimization problem; compare the models by defining a suitable goodness-of-fit or similarity measure; pick the model that best fits the data. Criteria for selecting models and the one that provides the best match are selected.

A number of criteria used to pick models for this analysis are stated below in Table 23.2. The interpretation of the criteria of selecting a model is higher the value of R^2 and $\bar{R}^2$, the better the fitness of the model. Also smaller the value of other criteria, viz., RMSE, AIC, BIC, MAE, and MAPE, the better is the fitness of the model.

Table 23.2 Computative forms – criteria for selecting model

Selection criterion	Notations/Abbreviations
$AIC = n \log (MSE) - 2k$	"AIC: Akaike Information Criterion"
$BIC = n \log (MSE) + K \log n$	"BIC: Bayesian Information Criterion "
$R^2 = 1 - \dfrac{Error\ Sum\ of\ Square}{Total\ sum\ of\ Square}$	"R^2: Coefficient of Determination" "$\bar{R}^2$: Adjusted Coefficient of Determination"
$\bar{R}^2 = 1 - (1 - R^2)\dfrac{n-1}{n-k}$	"MSE: Mean Squared Error" "RMS: Root Mean Square Error"
$MSE = \dfrac{1}{n-k}\sum \epsilon_t^2$	"MAE: Mean Absolute Error" "MAPE: Mean Absolute Percent Error"
$RMSE = \sqrt{\dfrac{1}{n-k}\sum \epsilon_t^2}$	'k' denotes the number of parameters in the model, n is the sample size,
$MAE = \dfrac{1}{n}\sum_{t=1}^n |\epsilon_t|$	"Y_t" represents observed value,
$MAPE = \dfrac{1}{n}\sum_{t=1}^n \left|\dfrac{\epsilon_t}{Y_t}\right| \times 100$	"ϵ_t: difference between the observed and estimated values."

3. Result and discussion

3.1 Description of milk production in India

During 1980–1981, the country produced 34.3 million tons of milk. Government-aided programs increased milk production during this time period. Milk output has been on the rise over the past three decades, as shown in Figure 23.1. The data indicate that milk production increased by 4.5% annually during the 11th five-year Plan (2007–2008 to 2011–2012). This has led to an increase in overall milk production during the time, from 107.9 million tones in 2007–2008 to 127.9 million tones in 2011–2012. Production of milk was at 132.4 million tons at the start of the 12th five-year Plan, or in 2012–2013. Milk output rose from 127.9 million tones in 2011–2012 to 132.4 million tones in 2012–2013, a rise of 3.5%. Lots of milk is produced in this nation, with a total of 187.75 million tones. The quantity of milk production increased by 6.5% compared to the previous year.

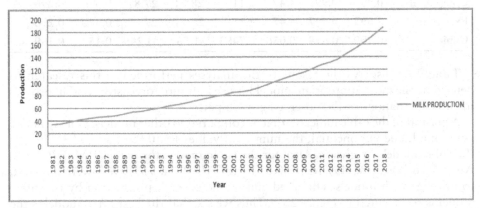

Figure 23.1 Milk production in India (1981–1982 to 2018–2019)

Table 23.3 Production of milk in India: Parameter estimates of growth models under study

Model	Parameters			
	a	*b*	*c*	*d*
Linear	16.976*	3.633*		
Quadratic	40.063*	0.200	0.089*	
Compound	34.358*	1.044*		
Exponential	34.358*	0.043*		
Cubic	29.382*	3.287*	-0.107*	0.003*

*indicates the statistical significance at the 5% level.

Table 23.3 displays the estimated value of the parameters of India's milk production for the data from 1981–1982 to 2018–2019. The analyses show that at the 5% level of significance, the constant parts of all five models are highly significant. Also, except for the quadratic model, the linear parts of all the models are also extremely significant. The quadratic and cubic parameter estimates of all the models are also remarkably significant. Under these conditions, it is very difficult and challenging to select the most appropriate model.

Thus, at this stage, following model selection criteria will be helpful for finding the best-fitted model. Table 23.4 exhibits the criteria for determining the best-fitted model for forecasting purposes and deciding growth patterns.

Table 23.4 Production of milk in India: Criteria for selecting model

Model	R^2	$\bar{R}^2$	MSE	RMSE	AIC	BIC	MAE	MAPE
Linear	0.939	0.937	32.82	5.7	61.61	60.75	1.18	1.490
Quadratic	0.992	0.991	5.04	2.244	32.69	31.40	0.470	0.63015
Compound	0.997	0.997	4.47	2.11	28.71	27.87	0.370	0.296
Exponential	0.997	0.997	4.47	2.11	28.71	27.87	0.370	0.296
Cubic	0.999	0.998	0.056	0.7483	-1.56	-3.249	0.152	0.189

Table 23.4 displays the results of the analyses performed in this research to determine the most appropriate model for use in future forecasting and in providing an explanation for the observed growth pattern.

A perusal of the table indicates that as compared to other models, namely linear, compound, and exponential, the numerical values R^2 (0.999) and $\bar{R}^2$ (.998) for the cubic model are greater. Also, for the cubic model, the values RMSE (0.748), AIC (-1.56), BIC (-3.249), MAE (0.152), and MAPE (0.189) are lower compared to other growth models. Observed and predicted milk production by the cubic model is represented in Figure 23.2. Thus we see that the best fitted model is the cubic model for illustrating the growth trend for production of milk in India and also for forecasting with the least error.

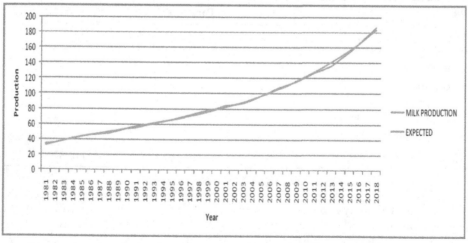

Figure 23.2 Observed and expected milk production in India (by cubic model) 2018–2019

3.2 Description of egg production in India

The nation produces 103.32 billion eggs annually. Egg production has increased by 8.5% in the last few years. The per capita annual supply of eggs in India is 79. Andhra Pradesh (19.1%), Tamil Nadu (18.2%), Telangana (13.2%), West Bengal (8.3%), and Haryana (5.7%) are the top five regions in terms of egg production. They are responsible for 65% of the country's overall egg output. The figure shows that the total egg production in the country was at 45.2 billion in the year 2004–2005, and since then, the production of eggs has continued to rise over the period. Figure 23.3 shows that egg output increased dramatically, peaking at 103.32 billion for the year 2018–2019.

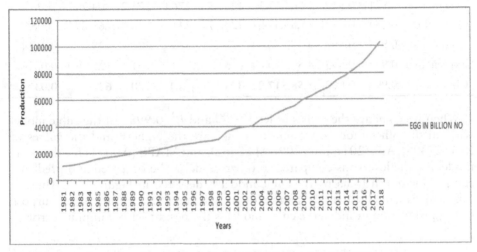

Figure 23.3 Egg production in India (1981–1982 to 2018–2019)

Table 23.5 Production of egg in India: Parameter estimates of growth models under study

Model	*a*	*b*	*c*	*d*
		Parameter		
Linear	-1506.521	2221.241*		
Quadratic	14,999.468*	-254.657	63.485*	
Compound	11,348.274*	1.059*		
Exponential	9600.971*	1305.978*		
Cubic	11,348.274*	0.057*	-35.268*	1.688*

*indicates the statistical significance at the 5% level.

Table 23.5 presents the estimated parameters of egg production in India for the years 1981–1982 to 2018–2019. Analyzing the table, we find that except for the linear model, the constant portion of all the models, quadratic, compound, exponential, and cubic, is highly significant. Also, except for the quadratic model, the linear portions of all other four models are notably significant at 5% level of significance. The remaining other quadratic and cubic coefficients of all the other models are also significant. Thus, for egg production, also it seems very difficult to select the best model.

So following diagnostic tools for model selection will be quite useful in determining the best-fitted model for describing the pattern of growth and forecasting egg production in India. The computed values of the model selection criteria are displayed in Table 23.6.

Table 23.6 Production of egg in India: Criteria for selecting model

Model	R^2	$\bar{R}^2$	MSE	RMSE	AIC	BIC	MAE	MAPE
Linear	0.921	0.919	144,47,774	1201.9	276.0	275.2	804.7	2.9
Quadratic	0.994	0.993	15,83,474.1	1258.71	241	240.3	269.89	15.54
Compound	0.995	0.995	4,57,830.64	676	223	201	129	0.0614
Exponential	0.995	0.995	4,57,830.64	676	223	201	129	0.0614
Cubic	0.998	0.998	1,58,517.02	398	201	200	62	0.0371

When we compare the values of R^2 (0.998) and $\bar{R}^2$ (0.998) of the cubic model with all the other models, we find that they are the highest, and the values of RMSE (398), AIC (201), BIC (200), MAE (62), and MAPE (0.0371) for the cubic model are the lowest as compared to other models. The observed and predicted egg production in India by the cubic model is depicted in Figure 23.4. Thus, in this case also, to elucidate the pattern of growth and also for forecasting purpose of egg production in India, the cubic model is the best fit with minimum error.

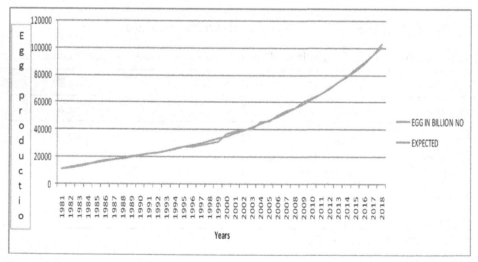

Figure 23.4 Observed and expected egg production in India (by cubic model) 2018–2019

Table 23.7 displays the production of milk and egg in India by the best fitted cubic model at 95% confidence interval for ten successive years. Predictions are made for the years 2019–2020 to 2028–2029. It is pertinent to mention here that the main drawback of the forecast is the forecasting error which increases with

Table 23.7 Milk and egg production in India (million tones): Forecast for ten successive years at 95% confidence interval.

Model	Description	Forecast year									
		2019–20	2020–21	2021–22	2022–23	2023–24	2024–25	2025–26	2026–27	2027–28	2028–29
Milk (cubic)	Lower limit	189.379	199.56	210.24	221.44	233.17	245.45	258.30	271.74	285.79	300.47
	Forecast	193.58	204.08	215.17	226.87	239.20	252.17	265.81	280.15	295.19	310.96
	Upper limit	197.79	208.61	220.11	232.307	245.22	258.89	273.33	288.55	304.58	321.45
	Length of CI	8.41	9.05	9.86	10.86	12.05	13.44	15.02	16.80	18.79	20.97
Egg (cubic)	Lower limit	1,04,288.83	1,10,502.45	1,16,994.03	1,23,770.15	1,30,838.76	1,38,208.77	1,45,889.61	1,53,890.89	1,62,222.21	1,70,893.05
	Forecast	1,07,026.37	1,13,448.07	1,20,204.37	1,27,305.41	1,34,761.31	1,42,582.19	1,50,778.20	1,59,359.44	1,68,336.07	1,77,718.2
	Upper limit	1,09,763.91	1,16,393.69	1,23,414.72	1,30,840.67	1,38,683.85	1,46,955.61	1,55,666.78	1,64,828.00	1,74,449.93	1,84,543.34
	Length of CI	5475.09	5891.23	6420.69	7070.53	7845.1	8746.84	9777.17	10,937.1	12,227.7	13,650.3

the period for which the forecast is made increases. Thus it is suggested that forecasts should be made for a short period of time rather than for a long period of time. It is seen from Table 23.7 the forecasting errors are sufficiently less, and the length of the confidence interval is also adequately small.

The research findings predicated that the production of milk and egg in India, for 2028–2029 would respectively be 310.96 million tons and 1,77,718.20 billion no if the present growth rate stayed same.

4. Conclusion

The aim of the time series analysis is to draw out useful statistics and other traits from time series data. Time series forecasting is predicting future values of time series using a model developed from past values. In the present study, five versions of deterministic time series models are examined. It meant that growth rates fluctuated wildly from year to year. The next ten years of expected milk and egg output in India were predicted to be 310.96083 million tons and 1,77,718.20166 billion numbers. The results of the study could help policymakers, researchers, and farmers make more precise short-term projections for national milk and egg production. As a policy implication, the present study would aid decision makers to take adequate measures to enhance the production of milk and eggs in India. Basically, reliable forecasts are dependent on precise information about the variables on which the models are based. Thus, the necessary requisite in order to get the best forecast results from forecasting models in our country is well-organized data warehouses.

References

1. Basic Animal Husbandry and Fisheries Statistics. (2019). Government of India.
2. Granger, C. W. T. and Newbold, P. (1986). Forecasting Economics Time Series. 2nd edition. Orlando, Florida: Academic Press Inc.
3. Gujarati, D. N. (2003). Basic Econometrics. 4th edition. New York: McGraw, Hill Inc., 23–26.
4. Hossain, M. J. and Hassan, M. F., (2013). Forecasting of milk, meat, and egg production in Bangladesh. *Res. J. Animal Vet. Fish. Sci.*, 1(9), 7–13.
5. Ministry of Fisheries. (2019). Animal Husbandry and Dairying 20th Livestock Census of India.
6. Pindyck, R. S. and Rubinfeld, D. L. (1998). Econometric models and Economic Forecasts. 4th edition. New York: McGraw, Hill Inc., 230–236.

24 Reliability analysis model for the consecutive *3-out-of-4*: F system

Sadhna Singh[a] and Amendra Singh[b]

Department of Mathematics, IBS, Khandari Campus, Dr. Bhimrao Ambedkar University, Agra-282002, India

Abstract

We have developed a reliability model for *3-out-of-4*: F with and its neural network. It is an ordered sequence with n components design in such a manner that failure of consecutive r components is a case of system failure. Further, we have determined the reliability of *2-out-of-3*: F system by using Laplace transformation.

Keywords: Reliability, *3-out-of-4*: F system, neural network, Laplace transformation

Introduction

The reliability model for *3-of-4*: F system is a very useful model in fault tolerance systems with several applications for decision, optimization, prediction, etc. The application may be conclusive or decision-based. In that system Fault can be defined in three ways: permanent fault, intermittent fault and transient fault. When a permanent fault occurs in the system then it has the continuous failure effects. The transient fault occurs in the system from external activities. The last intermittent fault occurs in the system from the design deficiency, and error within a component of a system. When the system experiences intermittent faults, then it affects the system at some instant of time and does not affect the system at another instant of time.

Habib and Szantai (2000) determined a new bound on r-out-of-n: F system. Habib and Youssef (2009) introduce the reliability analysis by using neural network model. Mikulin (2011) analyzed degradation models in reliability and survival analysis whereas Duer (2012) examined the dependability of any entity post regeneration within Artificial Neural Networks-based maintenance system. Radwan (2015) worked on *k-r-n*: F system and determined new bounds of increasing multi-state consecutive system while Amirian et al. (2020) found precision in reliability analysis for sequential multi-state events. Maohua (2022) modeled a combined zenith wet delay by applying neural network.

Modeling of Markov model for the consecutive *3-out-of-4*: F system

State (0): System will be in free state. **State (0:u_1):** system will be in the failure state with 1 minimal cut. **State (0:$u_{2(1)}$):** System will be in the failure state with ($2_{(1)}$) minimal cut. **State (0:$u_{2(2)}$):** System will be in the failure state with ($2_{(2)}$) minimal cut. **State (0:u_1):** One unit will be in the failure down state with 1

[a]sadhna20singh@gmail.com, [b]amendra1729@gmail.com

minimal cut. **State (1:u_1):** System will be in the failure down state with two units in the failure up state. **State (1:$u_{2(1)}$):** In this state, two units will be in failure up state and one unit in the failure down state with $(2_{(1)})$ minimal cuts. **State (1:$u_{2(2)}$):** In this state, two units in the failure up state and one unit in the failure down state with $(2_{(1)})$ minimal cuts. **State (1:1):** System will be in the failure up state with two units and one minimal cuts. **State (1:1; $u_{2(1)}$):** In this state, two units in the failure down state with 1 minimal cuts and one unit in the failure up state with $(2_{(1)})$ minimal cuts. **State (1:1; $u_{2(2)}$):** In this state, two units in the failure down state with 1 minimal cuts and one unit in the failure up state with $(2_{(1)})$ minimal cuts. **State (2:1):** System will be in the failure down state with three units and 1 minimal cut. **State (2:1; $u_{2(1)}$):** In this state two units in the failure down state with one minimal cuts and one unit in the failure up state with $(2_{(1)})$ minimal cuts. **State ($2_{(1)}$):** System will be in the failure state with $(2_{(1)})$ minimal cuts and one unit. **State ($2_{(1)}$:u_1):** In this state one unit in the failure down state with $(2_{(1)})$ minimal cuts and two units in the failure up state with 1 minimal cuts. **State ($2_{(1)}$:$u_{2(1)}$):** One unit will be in the failure down state with $(2_{(1)})$ minimal cuts **State (1:2$_{(1)}$):** System will be in the failure down state with two units and one minimal cuts. **State (1:2$_{(1)}$:$u_{2(1)}$):** In this state two units in the failure down state with $(2_{(1)})$ minimal cuts and one unit in the failure up state with $(2_{(1)})$ minimal cuts. **State (1:2$_{(1)}$):** System will be in the failure down state with three units and 1 minimal cuts. **State(2:1:2$_{(1)}$):** System will be in the failure down state and the state equations will be (Figures 24.1 and 24.2):

$$P_{t+\Delta t}(0) = \{1 - 4\lambda_\mu \Delta t - 4\lambda_d \Delta t\}P_t(0) + 3v\Delta t P_t(0, u_1) + v\Delta t P_t(0, u_{2(1)}) + v\Delta t P_t(0, u_{2(2)}) \tag{1}$$

$$P_{t+\Delta t}(0:u_1) = \{1 - 3v\Delta t\}P_t(0, u_1) + 3\lambda_\mu \Delta t P_t(0) \tag{2}$$

$$P_{t+\Delta t}(0:u_{2(1)}) = \{1 - v\Delta t\}P_t(0, u_{2(1)}) + \lambda_\mu \Delta t P_t(0) \tag{3}$$

$$P_{t+\Delta t}(0:u_{2(2)}) = \{1 - v\Delta t\}P_t(0, u_{2(2)}) + \lambda_\mu \Delta t P_t(0) \tag{4}$$

$$P_{t+\Delta t}(1) = \{1 - 3\lambda_\mu \Delta t - 3\lambda_d \Delta t\}P_t(1) + v\Delta t P_t(1, u_1) + v\Delta t P_t(1, u_{2(1)}) + v\Delta t P_t(1, u_{2(2)}) + 3\lambda_d \Delta t P_t(0) \tag{5}$$

$$P_{t+\Delta t}(1:u_1) = \{1 - 3v\Delta t\}P_t(1, u_1) + 3\lambda_\mu \Delta t P_t(1) \tag{6}$$

$$P_{t+\Delta t}(1:u_{2(1)}) = \{1 - v\Delta t\}P_t(1, u_{2(1)}) + \lambda_\mu \Delta t P_t(1) \tag{7}$$

$$P_{t+\Delta t}(1:u_{2(2)}) = \{1 - v\Delta t\}P_t(1, u_{2(2)}) + \lambda_\mu \Delta t P_t(1) \tag{8}$$

$$P_{t+\Delta t}(1;1) = \{1 - 2\lambda_\mu \Delta t - 2\lambda_d \Delta t\}P_t(1;1) + v\Delta t P_t(1;1;u_{2(1)}) + v\Delta t P_t(1;1;u_{2(2)}) + 2\lambda_d \Delta t P_t(1) \tag{9}$$

$$P_{t+\Delta t}(1;1;u_{2(1)}) = \{1 - v\Delta t\}P_t(1;1;u_{2(1)}) + \lambda_\mu \Delta t P_t(1;1) \tag{10}$$

$$P_{t+\Delta t}(1;1;u_{2(2)}) = \{1 - v\Delta t\}P_t(1;1;u_{2(2)}) + \lambda_\mu \Delta t P_t(1;1) \tag{11}$$

$$P_{t+\Delta t}(2;1) = \{1 - \lambda_\mu \Delta t - \lambda_d \Delta t\}P_t(2;1) + v\Delta t P_t\big(2;1;u_{2(2)}\big) + \lambda_d \Delta t P_t(1;1) \qquad (12)$$

$$P_{t+\Delta t}\big(2_{(1)}\big) = \{1 - 3\lambda_\mu \Delta t - 3\lambda_d \Delta t\}P_t\big(2_{(1)}\big) + 3v\Delta t P_t\big(2_{(1)};u_{(1)}\big) + v\Delta t P_t\big(2_{(1)},u_{2(2)}\big) + \lambda_d \Delta t P_t(0) \qquad (13)$$

$$P_{t+\Delta t}\big(2_{(1)}:u_1\big) = \{1 - 3v\Delta t\}P_t\big(2_{(1)};u_1\big) + 3\lambda_\mu \Delta t P_t\big(2_{(1)}\big) \qquad (14)$$

$$P_{t+\Delta t}\big(2_{(1)}:u_{2(1)}\big) = \{1 - v\Delta t\}P_t\big(2_{(1)};u_{2(1)}\big) + \lambda_\mu \Delta t P_t\big(2_{(1)}\big) \qquad (15)$$

$$P_{t+\Delta t}\big(1;2_{(1)}\big) = \{1 - 2\lambda_\mu \Delta t - 2\lambda_d \Delta t\}P_t\big(1;2_{(1)}\big) + v\Delta t P_t\big(1;2_{(1)},u_{(1)}\big) + 3\lambda_d \Delta t P_t\big(2_{(1)}\big) + \lambda_d \Delta t P_t(1) \qquad (16)$$

$$P_{t+\Delta t}\big(1;2_{(1)}:u_1\big) = \{1 - v\Delta t\}P_t\big(1;2_{(1)},u_1\big) + \lambda_\mu \Delta t P_t\big(1;2_{(1)}\big) \qquad (17)$$

$$P_{t+\Delta t}\big(1;1;2_{(1)}\big) = \{1 - \lambda_\mu \Delta t - \lambda_d \Delta t\}P_t\big(1;1;2_{(1)}\big) + \lambda_d \Delta t P_t(1;1) + \lambda_d \Delta t P_t\big(1;2_{(1)}\big) \qquad (18)$$

$$P_{t+\Delta t}\big(2;1;2_{(1)}\big) = P_t\big(2;1;2_{(1)}\big) + \lambda_d \Delta t P_t(1;1) + \lambda_d \Delta t P_t(2;1) \qquad (19)$$

$$P_{t+\Delta t}\big(2;1;u_{2(2)}\big) = \{1 - v\Delta t\}P_{t+\Delta t}\big(2;1:u_{2(2)}\big) + \lambda_\mu \Delta t P_t(2;1) \qquad (20)$$

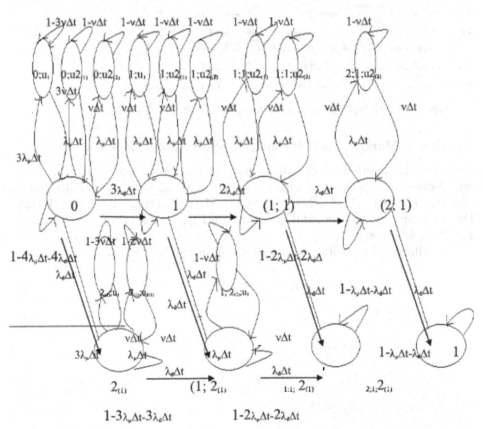

Figure 24.1 The discrete time Markov model of a consecutive *3-out-of-4*: F system

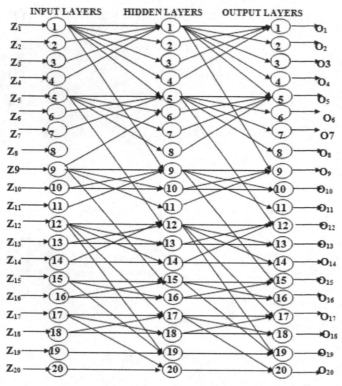

Figure 24.2 Neural network of a consecutive 3-*out-of-*4: F system

Solution of Markov model for the consecutive 2-*out-of-*3: F system

Now for the solution of the above type of equations, taking a smaller system 2-*out-of* 3: F system in which there are only 12 state namely (0), (0;*u*1), (0;*u*21), (1), (1;*u*1), (1;*u*2), (21), (21;*u*1), (21;*u*1) (1;1), (1;1;*u*21), (1;21), (1;1;21). For the simplicity we denoted these state with $P_1, P_2, P_3, P_4, P_5, P_6, P_7, P_8, P_9, P_{10}, P_{11}, P_{12}$, respectively.

The differential equations for 2-*out-of-*3: F system is as follows:

$$P_1'(t) = -3(\lambda_\mu + \lambda_d)P_1(t) + 2vP_2(t) + vP_3(t) \tag{21}$$

$$P_2'(t) = 2\lambda_\mu P_1(t) - 2vP_2(t) \tag{22}$$

$$P_3'(t) = \lambda_\mu P_1(t) - vP_3(t) \tag{23}$$

$$P_4'(t) = -2(\lambda_\mu + \lambda_d)P_1(t) - vP_5(t) + vP_6(t) + 2\lambda_d P_1(t) \tag{24}$$

$$P_5'(t) = \lambda_\mu P_4(t) - vP_5(t) \tag{25}$$

$$P_6'(t) = \lambda_\mu P_4(t) - vP_6(t) \tag{26}$$

$$P_7'(t) = -2(\lambda_\mu + \lambda_d)P_7(t) + 2vP_8(t) + \lambda_d P_1(t) \tag{27}$$

$$P_8'(t) = 2\lambda_\mu P_7(t) - 2vP_8(t) \tag{28}$$

$$P_9'(t) = -(\lambda_\mu + \lambda_d)P_9(t) + vP_{10}(t) + \lambda_d P_4(t) \tag{29}$$

$$P_{10}'(t) = \lambda_\mu P_9(t) - vP_{10}(t) \tag{30}$$

$$P_{11}'(t) = 2\lambda_d P_9(t) + \lambda_d P_4(t) \tag{31}$$

$$P_{12}'(t) = \lambda_d P_9(t) \tag{32}$$

Now solving these equations using Laplace transformation, obtained P_{12} for the reliability of the system.

$$L_{12}(s) = [2\lambda_d^3(s + 2v)(s + v)^2] \times \left[s^3 + s^2(3\lambda_\mu + 3\lambda_d + 3v) + sv(6\lambda_\mu + 9\lambda_d + 2v) + v^2(4\lambda_\mu + 6\lambda_d) + \right.$$
$$\left. s(s + 2\lambda_\mu + 2\lambda_d)[(s + v)(s + 2\lambda_\mu + 2\lambda_d) - \lambda_\mu v]\right]^{-1} \tag{33}$$

Then the system reliability will be

$$L[R_s(t)] = L[1 - P_{12}(t)] \tag{34}$$

$$L[R_s(t)] = s - L\left[[2\lambda_d^3(s + 2v)(s + v)^2] \times \left[s^3 + s^2(3\lambda_\mu + 3\lambda_d + 3v) + sv(6\lambda_\mu + 9\lambda_d + 2v) + \right.\right.$$
$$\left.\left. v^2(4\lambda_\mu + 6\lambda_d) + s(s + 2\lambda_\mu + 2\lambda_d)[(s + v)(s + 2\lambda_\mu + 2\lambda_d) - \lambda_\mu v]\right]^{-1}\right] \tag{35}$$

The desired reliability can be obtained by taking the inverse Laplace transform of (35).

Conclusion

In this paper, we have formulated a reliability model for the consecutive *3-out-of-4*: F system by using neural network techniques. Further, we have determined the reliability of *2-out-of-3*: F system by using Laplace transform. These types of system are used in many areas such as telecommunication system, oil pipeline and digital computer systems.

References

1. Manzoul, M. A. and Suliman, M. (1990). Neural network for reliability analysis of simplex systems. *Microelec. Reliab.*, 30(4), 795–800.
2. Habib, A. and Szantai, T. (2000). New bounds on the reliability of the consecutive k-out-of-r-from-n: F system. *Reliab. Engg. Sys. Safety,* 68, 97–104.
3. Habib, A., Alsieidi, R., and Youssef, G. (2009). Reliability analysis of a consecutive r-out-of- n: F system based on neural networks. *Chaos Solitons Fract.*, 39, 610–624.
4. Mikulin, N. (2011). Degradation models in reliability and survival analysis. *Int. En-cyclop. Stat. Sci.*, 354–363.
5. Duer, S. (2012). Examine of reliability of a technical object after it's regeneration in a maintenance system with an artificial neural networks. *Adv. Space Res.*, 21(3), 523–534.

6.	Radwan, T. (2015). New bounds of increasing multi-state consecutive k-out-of-r-from-n: F system by applying the third order Boole-Bonferroni bounds. *Appl. Math. Sci.*, 9(109), 5417–5428.

7.	Amirian, Y., Khodadadi, A., and Chatrabgoun, O. (2020). Exact reliability for a multi-state consecutive linear (circular) k -out-of- r -from- n: F system. *Int. J. Reliab. Qual. Safety Engg.*, 27(1), 2050003 (26 pages).

8.	Maohua, D. (2022). Developing a new combined model of zenith wet delay by using neural network. *Adv. Space Res.*, 70(2), 350–359.

25 Fuzzy inference-based system for the authenticity of a claim raised by the insurer

Monika Rathore[1,a], N. K. Sharma[2] and Aryaman Kumar Sharma[3]

[1]Department of Mathematics, C. L. Jain College, Firozabad, Dr. Bhimrao Ambedkar University, Agra, Uttar Pradesh, India

[2]Department of Computer Science, IIMT, Knowledge Park, Greater Noida, Uttar Pradesh, India

[3]Department of Electronics and Communication Engineering, Jaypee Institute of Information Technology, Noida, Uttar Pradesh, India

Abstract

Nowadays insurers perform their work through conventional method on the basis of the insurer's former practices and surveys done by them. Here some more factors also exist, but the conventional method always assists holders those who own negative credits but hinder those holders who own the positive credits. Thus, there is a need to have a suitable premium of the insurance that would be beneficial for holders who own positive credits and should reward with a bonus. While, holders should get malus who own negative credits. Further, it is observed that several times the claimant files a vague record and irrelevant information while making a claim. Therefore, it is very important to have a model for such type of vague or partial information. Thus, this proposed work is designed to examine the authenticity of the claim raised by the insurer by using the fuzzy logic system. The variables used in the proposed fuzzy inference system are the vagueness index, degree of incompleteness, level of judgment, and credit report of the claimant while authentic settlement and settlement with fraud are taken as output.

Keywords: Insurance, risk, fuzzy logic, authenticity, triangular membership function, trapezoidal membership function

Introduction

In insurance, the "Bonus Malus" is a system that balances the premium paid by a holder according to his claim history. Generally, the bonus is a discount on the premium which is specified on the renewal of the policy if any claim is not made in the preceding year while malus is an increment in the premium if the claim is done in the preceding year. Normally, bonus–malus systems are very ordinary in the insurance of vehicles. In Britain and Australia, this arrangement is known as a "No-Claim Discount" (NCD) or "No-Claim Bonus" (NCB). This system in insurance files good and bad events in the history of the holder to determine his premium in the present. Premiums increase when holders have a circumstance

[a]monikarathore16sep@gmail.com, [b]naveen.sharma@iilm.ac.in, [c]kumararyaman26@gmail.com

that is "bad" likewise avoiding claims is "good" and will reduce the holder's premium. In many standards, insurance agreements are based on the bonus–malus system.

Literature review

Several studies have shown that in today's world, insurance policies have become widespread. In this process, Kumar and Pathak (2009) provided a third dimension in place of the existing dynamic method to calculate premiums. They used a fuzzy logic approach to determine the premium. Abdullah and Rahman (2012) proposed a study by utilizing the fuzzy logic for comparing likelihood to purchase insurance of health on the basis of some variables. Kumar and Jain (2012) presented a model by using the fuzzy inference engine to assist the diabetic holders for life insurance. Kumar and Tiwari (2015) designed a model to assess the risk of cancellation of policies by using the concept of a fuzzy inference system. Hooda and Kumari (2017) presented the concept of hybridization of fuzzy sets. They also proposed a new approach for converting table of soft set to the table of fuzzy soft set. Kumar et al. (2017) proposed a study, by applying the fuzzy logic to provide symptomatic output of the cancellation risk in the future. Kalra et al. (2022) carried out a fuzzy expert system to recognize medical insurance frauds in medical coverage and also in all areas of insurance.

Key features

This proposed system makes use of fuzzy logic expert system to identify the authenticity of the claim which is outlined around the concept of "index of vagueness". The "index of vagueness" has been an important tool to deal with human reasoning and also in the process of rough calculation, uncertainty, and vagueness. Fuzzy logic assesses the information of the related set to different variables in the form of membership functions.

$$X_1 = \frac{\left(\sum_{i=1}^{I} \sum_{j=1}^{J} W_{ij}\Delta_{ij}\right)}{I} \tag{1}$$

In fuzzy logic the concept of assigning weights to different variables basically in the study of settling insurance claim proved as an important step to obtain optimum output. Weighted decision-making is a common approach in the fields including finance and risk assessment and also have a higher impact on the decision-making process, and the weights for a certain set of information sum up to one (unity).

Methodology

In the proposed work, a triangular membership function is used for the input variable "degree of incompleteness" while trapezoidal membership functions are used for the input variable "vagueness index", "level of judgment", and "credit report of the claimant" while setting the claim, need to be verified to reduce the turnaround time of claim settlements.

Algorithm (using fuzzy logic)

1) **Input:** First the crisp values of the input factors – degree of incompleteness, vagueness index, level of judgment, and credit report of the claimant are collected through various sources such as forms filled by insurant or holder, etc.

2) **Comparison with a minimal value:** If the settlement of a claim is smaller than a preset digit, which is determined by the auditor, then the settlement of the claim is to be considered authentic settlement and then go to step 8.

3) **Evaluate the input factors:** Determine the input factors – vagueness index x_1 degree of incompleteness x_2, level of judgment x_3, and credit report of the claimant x_4 (Figure 25.1).

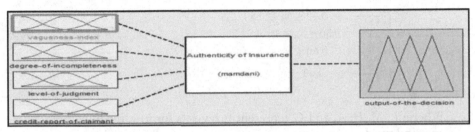

Figure 25.1 FIS for authentication

4) **Fuzzify the inputs:** The equations of triangle and trapezoidal membership functions of all input factors are as follow (Figures 25.2–25.5):

$$\mu_L(x_1) = \begin{cases} 1 & x_1 \le 0.35 \\ \frac{0.45-x_1}{0.10} & 0.35 < x_1 \le 0.45 \end{cases}$$

$$\mu_M(x_1) = \begin{cases} \frac{x_1-0.35}{0.10} & 0.35 \le x_1 < 0.45 \\ 1 & 0.45 \le x_1 < 0.55 \\ \frac{0.65-x_1}{0.10} & 0.55 \le x_1 < 0.65 \end{cases} \tag{2}$$

$$\mu_H(x_1) = \begin{cases} \frac{x_1-0.55}{0.10} & 0.55 \le x_1 < 0.65 \\ 1 & x_1 \ge 0.65 \end{cases}$$

$$\mu_L(x_2) = \begin{cases} \frac{x_2}{0.23} & x_2 \le 0.23 \\ \frac{0.45-x_2}{0.22} & 0.23 \le x_2 \end{cases}$$

$$\mu_M(x_2) = \begin{cases} \frac{x_2-0.35}{0.20} & x_2 < 0.55 \\ \frac{0.75-x_2}{0.20} & 0.55 \le x_2 \end{cases} \tag{3}$$

$$\mu_H(x_2) = \begin{cases} \frac{x_2-0.55}{0.21} & x_2 < 0.76 \\ \frac{1-x_2}{0.20} & 0.76 \le x_2 \end{cases}$$

$$\mu_L(x_3) = \begin{cases} \frac{x_3-0}{0.15} & 0 \le x_3 < 0.15 \\ 1 & 0.15 \le x_3 < 0.30 \\ \frac{0.45-x_3}{0.15} & 0.30 \le x_3 < 0.45 \end{cases}$$

$$\mu_M(x_3) = \begin{cases} \frac{x_3-0.30}{0.15} & 0.30 \le x_3 < 0.45 \\ 1 & 0.45 \le x_3 < 0.60 \\ \frac{0.75-x_3}{0.15} & 0.60 \le x_3 < 0.75 \end{cases}$$

$$\mu_H(x_3) = \begin{cases} \frac{x_3-0.60}{0.15} & 0.60 \le x_3 < 0.75 \\ 1 & x_3 \ge 0.75 \end{cases}$$

(4)

$$\mu_L(x_4) = \begin{cases} 1 & x_4 \le 0.40 \\ \frac{0.50-x_4}{0.10} & 0.40 \le x_4 < 0.50 \end{cases}$$

$$\mu_A(x_4) = \begin{cases} \frac{x_4-0.40}{0.10} & 0.40 \le x_4 < 0.50 \\ 1 & 0.50 \le x_4 < 0.60 \\ \frac{0.70-x_4}{0.10} & 0.60 \le x_4 < 0.70 \end{cases}$$

$$\mu_G(x_4) = \begin{cases} \frac{x_4-0.60}{0.10} & 0.60 \le x_4 < 0.70 \\ 1 & x_4 \ge 0.70 \end{cases}$$

(5)

Where: *L, M, H, A*, and *G* represent fuzzy sets for low, medium, high, average, and good.

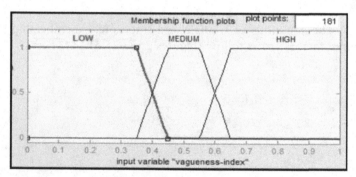

Figure 25.2 Membership function plot of vagueness index

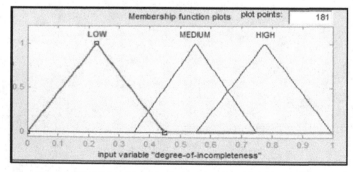

Figure 25.3 Membership function plot of degree of incompleteness

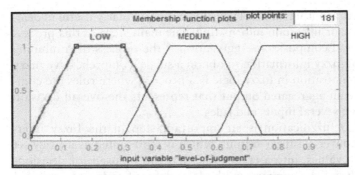

Figure 25.4 Membership function plot of level of judgment

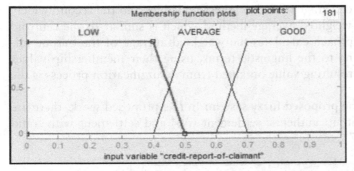

Figure 25.5 Membership function plot of credit report of the claimant

5) **Rule base corresponding to input factors:** Fuzzy logic system is generally based on conditional statement. A conditional statement, often denoted as *"if-then"* or *"if ... then"* statement, which consists of two parts: antecedent (premise) *"if"* and consequence (conclusion) *"then"*. The *"if"* part of the statement represents the condition or situation that needs to be true in order for something else to happen. While, the *"then"* part of the statement represents the result or action that follows if the antecedent is true. Here in this proposed work, three inputs are classified into three fuzzy sets which are defined as – Low, Medium, and High and the fourth fuzzy set is defined as – Low, Average, and Good. Therefore, 81 (3×3×3×3) fuzzy rule-base are to be formed here. In this proposed system there are two outputs: authentic settlement (AS) and settlement with fraud (SF) (Figure 25.6).

```
1. If (vagueness-index is LOW) and (degree-of-incompleteness is LOW) and (level-of-judgment is LOW) and (credit-report-of-claimant is LOW) then (output-of-the-decision is SETTLEMENT-OF-FRAUD) (1)
2. If (vagueness-index is LOW) and (degree-of-incompleteness is LOW) and (level-of-judgment is LOW) and (credit-report-of-claimant is AVERAGE) then (output-of-the-decision is SETTLEMENT-OF-FRAUD) (1)
3. If (vagueness-index is LOW) and (degree-of-incompleteness is LOW) and (level-of-judgment is LOW) and (credit-report-of-claimant is GOOD) then (output-of-the-decision is AUTHENTIC-SETTLEMENT) (1)
4. If (vagueness-index is LOW) and (degree-of-incompleteness is LOW) and (level-of-judgment is MEDIUM) and (credit-report-of-claimant is LOW) then (output-of-the-decision is AUTHENTIC-SETTLEMENT) (1)
5. If (vagueness-index is LOW) and (degree-of-incompleteness is LOW) and (level-of-judgment is MEDIUM) and (credit-report-of-claimant is AVERAGE) then (output-of-the-decision is AUTHENTIC-SETTLEMENT) (1)
6. If (vagueness-index is LOW) and (degree-of-incompleteness is LOW) and (level-of-judgment is MEDIUM) and (credit-report-of-claimant is GOOD) then (output-of-the-decision is AUTHENTIC-SETTLEMENT) (1)
7. If (vagueness-index is LOW) and (degree-of-incompleteness is LOW) and (level-of-judgment is HIGH) and (credit-report-of-claimant is LOW) then (output-of-the-decision is AUTHENTIC-SETTLEMENT) (1)
8. If (vagueness-index is LOW) and (degree-of-incompleteness is LOW) and (level-of-judgment is HIGH) and (credit-report-of-claimant is AVERAGE) then (output-of-the-decision is AUTHENTIC-SETTLEMENT) (1)
9. If (vagueness-index is LOW) and (degree-of-incompleteness is LOW) and (level-of-judgment is HIGH) and (credit-report-of-claimant is GOOD) then (output-of-the-decision is AUTHENTIC-SETTLEMENT) (1)
10. If (vagueness-index is LOW) and (degree-of-incompleteness is MEDIUM) and (level-of-judgment is LOW) and (credit-report-of-claimant is LOW) then (output-of-the-decision is SETTLEMENT-OF-FRAUD) (1)
11. If (vagueness-index is LOW) and (degree-of-incompleteness is MEDIUM) and (level-of-judgment is MEDIUM) and (credit-report-of-claimant is AVERAGE) then (output-of-the-decision is AUTHENTIC-SETTLEMENT) (1)
```

Figure 25.6 Rule base for authenticity of insurance

6) **Initiating the process of fuzzy inference system:** Initiating a fuzzy inference engine involves fuzzy logic to process input variables through a set of fuzzy

rule-base to generate output values. Fuzzy logic is particularly useful in dealing with imprecise or uncertain information. The main steps in this process are aggregation and composition. Aggregation is the process of combining multiple pieces of fuzzy information to obtain a single, comprehensive fuzzy set. While, the composition in fuzzy logic is a process, where rules are combined to generate an aggregated output that represents the overall decision or action based on several inputs and rules.

7) **Defuzzification:** Defuzzification is an important step in the fuzzy inference process. It converts a fuzzy output, which is represented as a fuzzy set with membership values, into a crisp value that can be used for decision-making and analysis. The common methods of defuzzification are centroid of area, which calculates the center of gravity of membership function to find the weighted average of the membership values and the second widely used method is weighted average method which is similar to the centroid method, this approach calculates the weighted average of the output values corresponding to the linguistic terms, using their membership values as weights. The resulting value obtained from defuzzification process is the crisp output.

8) **The output of the proposed fuzzy system:** In this proposed work, there are two types of output: authentic settlement (AS) and settlement with fraud (FS) (Figure 25.7).

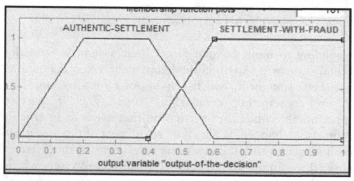

Figure 25.7 Membership function plot of the output of the decision

$$\mu_{AS}(x_5) = \begin{cases} \dfrac{x_5 - 0}{0.20} & 0 \leq x_5 < 0.20 \\ 1 & 0.20 \leq x_5 < 0.40 \\ \dfrac{0.60 - x_5}{0.20} & 0.40 \leq x_5 < 0.60 \end{cases}$$

$$\mu_{SF}(x_5) = \begin{cases} \dfrac{x_5 - 0.40}{0.20} & 0.40 \leq x_5 < 0.60 \\ 1 & x_5 \geq 0.60 \end{cases}$$

(6)

Where AS, and SF represent fuzzy sets for authentic settlement, and settlement with fraud.

Case study

For an illustration, here insurer make use of these four input variables – vague-ness index x_1, degree of incompleteness x_2. level of judgment x_3, and the credit report of claimant x_4. These indices represent the authentic settlement of an assur-ance claim settlement and these inputs also represent the degree of vagueness present in the information additionally level of judgment is used to measure the unauthentic claim in the insurance.

1. **Input:** Let the claim value = Rs. 80,000.00
2. **Comparison with a threshold:** If values are greater than a prefixed value (let Rs. 25,000.00). Then the authenticity of the claim settlement is compared by the proposed fuzzy system.
3. **Evaluate the authenticity:** In this step the input values of the claim settlement are evaluated. Let the values of inputs be: $x_1 = 0.38$, $x_2 = 0.70$, $x_2 = 0.65$, $x_4 = 0.62$ (Say)
4. **Fuzzification of input values:** Using the triangular and trapezoidal member-ship functions, the degree of membership functions of each fuzzy set after putting values from step-3 is as follow:

$\mu_L(x_1) = \frac{0.45-x_1}{0.10} = 0.70 \qquad \mu_M(x_1) = \frac{x_1-0.35}{0.10} = 0.30 \qquad \mu_H(x_1) = 0$

$\mu_L(x_2) = \max\left\{0, \frac{0.45-x_2}{0.22}\right\} = 0 \quad \mu_M(x_2) = \max\left\{0, \frac{0.75-x_2}{0.20}\right\} = 0.25 \quad \mu_H(x_2) = \max\left\{0, \frac{x_2-0.55}{0.21}\right\} = 0.71$

$\mu_L(x_3) = 0 \qquad \mu_M(x_3) = \frac{0.75-x_3}{0.15} = 0.67 \qquad \mu_H(x_3) = \frac{x_3-0.60}{0.15} = 0.33$

$\mu_L(x_4) = 0 \qquad \mu_A(x_4) = \frac{0.70-x_4}{0.10} = 0.80 \qquad \mu_G(x_4) = \frac{x_4-0.60}{0.10} = 0.20$

5. **Fire rule bases:** Rule bases under consideration to four input variables on the basis of the fuzzy membership functions are as follow:
    Rule 14: if x_1 is LOW, x_2 is MEDIUM, x_3 is MEDIUM, and x_4 is AVERAGE then Y is an AUTHENTIC SETTLEMENT.
    Rule 15: if x_1 is LOW, x_2 is MEDIUM, x_3 is MEDIUM, and x_4 is GOOD then Y is an AUTHENTIC SETTLEMENT.
    Rule 23: if x_1 is LOW, x_2 is HIGH, x_3 is MEDIUM, and x_4 is AVERAGE then Y is a SELLTLEMENT WITH FRAUD.
    Rule 26: if x_1 is LOW, x_2 is HIGH, x_3 is HIGH, and x_4 is AVERAGE then Y is a SETTLEMENT WITH FRAUD.
    Rule 41: if x_1 is MEDIUM, x_2 is MEDIUM, x_3 is MEDIUM, and x_4 is AVERAGE then Y is an AUTHENTIC SETTLEMENT.
    Rule 44: if x_1 is MEDIUM, x_2 is MEDIUM, x_3 is HIGH, and x_4 is AVERAGE then Y is a SETTLEMENT WITH FRAUD.
    Rule 45: if x_1 is MEDIUM, x_2 is MEDIUM, x_3 is HIGH, and x_4 is GOOD then Y is an AUTHENTIC SETTLEMENT.
    Rule 50: if x_1 is MEDIUM, x_2 is HIGH, x_3 is MEDIUM, and x_4 is AVERAGE then Y is a SETTLEMENT WITH FRAUD.

Rule 54: if x_1 is MEDIUM, x_2 is HIGH, x_3 is HIGH, and x_4 is GOOD then Y is a SETTLEMENT WITH FRAUD.

6.   **Execution of the inference engine:** In this proposed fuzzy inference system, the "root sum square" method is used to merge the conclusion of all rules, and the function are also scaled at their corresponding magnitude. Generally, the range of respective output membership function strength is [0–1]:

$$Authentic\ Settlement = \sqrt{\Sigma_{i \in AS}(\mu_{R_i})^2}$$

(7)

$$Authentic\ Settlement = 0.53$$

$$Settlement\ with\ fraud = \sqrt{\Sigma_{i \in SF}(\mu_{R_i})^2}$$

(8)

$$Settlement\ with\ Fraud = 0.998$$

7.   **Defuzzification:** In this work, weighted average method is used for defuzzification process. The weighted average method is commonly employed in the defuzzification step of a fuzzy logic-based system to make decisions based on imprecise input data. The weighted average method is calculated by multiplying the membership values of each output set with their corresponding weights and then sum up these weighted values. At last, divide the sum by the total sum of weights. The output of this proposed fuzzy expert system for this case study is 0.857, which belongs to the set of settlements with fraud (SF) more than the set of authentic settlements (AS).

8.   **Output of the proposed fuzzy system:** Regarding the case being reviewed, the system's output shows that the case is of fraud settlement. So, the insurance company will give only the threshold amount, i.e., the amount which is predetermined by the insurance company for a particular case. So, the insurance company will give only Rs. 25,000.00 to the claimant.

Conclusion

The proposed work presents a fuzzy rule-based model to evaluate the indicative results of the authenticity of a claim using four prominent input factors. This work is limited to these four input factors those have significant impact in the realm of insurance sector. This study introduces utilization of the fuzzy inference system in conjunction with MATLAB, offers an optimal platform for representing a variety of combinations processed from a subset of the tried combinations.

However, it should be noted that this method is not suitable to detect fraudulent activities during the process of insurance claims. Previous works in this domain has presented another method in which neural network-based technique is used to identify the occurrence of management fraud in financial datasets. In this work, a fuzzy-based expert system has been proposed specifically to detect fraud in insurance claims. The forthcoming endeavors of this proposed work will focus on modifying the system's performance by adjusting input functions. Additionally, possible alternative for future expansion involves- the duration to settle up the claim, higher authorities consulted in the settlement process, and the frequency of similar claims submitted by the claimant. In conclusion, it would be fair to say that including these factors in the system's rule base can enhance its capabilities, but it is important to manage the complexity involve in results from real-life scenario. Hence, this approach aims to obtain optimal performance while adapting the complexities of real-world situations. The fuzzy inference system proposed in this study is subjected to do evaluation using the hypothetical data.

References

1. Zadeh, L. A. (1965). Fuzzy sets. *Inform. Control,* 8(3), 338–353.
2. Dewit, G. W. (1982). Underwriting and uncertainty. *Ins. Math. Econom.,* 1(4), 277–285.
3. Buckley, J. J. (1987). The fuzzy mathematics of finance. *Fuzzy Sets Sys.,* 21(3), 257–273.
4. Buckley, J. J. (1992). Solving fuzzy equations in economics and finance. *Fuzzy Sets Sys.,* 48(3), 289–296.
5. Deshmukh, A. and Laxminarayan, T., (1998). A rule-based fuzzy reasoning system for assessing the risk of management fraud. *Int. Sys. Acc. Fin. Manag.,* 7(4), 223–241.
6. Kumar, S. and Pathak, P. (2009). Premium allocation- fuzzy approach in insurance business. *Proc. 3rd National Conf. INDIACom,* 703–705.
7. Abdullah, L. and Rahman, N. A. (2012). Employee likelihood of purchasing health insurance using fuzzy inference system. *Int. J. Comp. Sci. Iss.,* 9(1), 112–116.
8. Kumar, S. and Jain, H. (2012). A fuzzy logic based model for life insurance underwriting when insurer is diabetic. *Eur. J. Appl. Sci.,* 4(5), 196–202.
9. Kumar, S. and Tiwari, N. (2015). Mathematical model for the risk of cancellation of life insurance policies. *SRM Int. J. Engg. Sci.,* 3(1), 19–26.
10. Hooda, D. S. and Kumari, R. (2017). On application of fuzzy soft sets in dimension reduction and medical diagnosis. *Adv. Res.,* 12(2), 1–9.
11. Kumar, S., Chaudhary, S., and Sharma, G. (2017). A fuzzy logic approach to calculate the risk of cancellation of policies. *Agra University J. Res. Sci.,* 1(2), 1–8.
12. Kalra, G., Rajoria, Y. K., Boadh, R., Rajendra, P., Pandey, P. Khatak, N., and Kumar, A. (2022). Study of fuzzy expert systems towards prediction and detection of fraud case in health care insurance. *Mat. Today Proc.,* 56(1), 477–480.

26 The use of fuzzy inference system as a method for regulating the rate of human population rise

Anurag Paliwal[1,a] and Satyendra Singh Yadav[2,b]

[1]Research Scholar, Narain College, Shikohabad, Distt-firozabad, Affiliated at Dr. Bhimrao Ambedkar University Agra, Uttar Pradesh, India
[2]Professor, Narain College, Shikohabad, Distt-firozabad, Affiliated at Dr. Bhimrao Ambedkar University Agra, Uttar Pradesh, India

Abstract

An achievable modeling and simulation methodology for estimating population growth across regions is presented, and In order to accomplish this objective, this paper presents an introduction to the concept of fuzzy inference. Traditional techniques of growth modeling and prediction are complex and time intensive, which prompted the current investigation. Following a presentation of the relevant design difficulties, it is decided to construct the fuzzy inference model for the rise in population. The criteria utilized in traditional methods of population forecasting are based on vague assumptions about human society and the economy. They are then fed into an inherently fuzzy model of population growth based on fuzzy inferences, which provides an estimate of the growth rate. In order to test the fuzzy population model, we use a fuzzy inference system (FIS) that has already been developed.

Keywords: Fuzzy inference system, Linguistic variables, Membership function, Population growth, Defuzzification

1. Introduction

Ecologists studying populations focus on factors including births, deaths, migration, and immigration that affect population growth and decline. Most ecological population growth models developed over the years use the concept of a stable equilibrium level around which population size fluctuations occur over a given time period. Due to the presence of numerous environmental resources, the values of the parameters that affect this growth shape remain unknown. It's hard to get an accurate read on the quantity of natural resources in a highly populated area. It is clear that fuzzy logic (FL), which was first developed to describe fuzziness and uncertainty in the actual world, has a significant impact on the evolution of ecological management issues. FL enables the use of fuzzy rules in the evaluation of the parameters affecting population growth.

Taking into consideration the degree of subjectivity inherent in either the state variables or the parameters, Barros et al. (2000) conduct an analysis of the behavior of models that represent the dynamics of a population. As Diab and Saade (2005) shown, fuzzy inference may be used to create a realistic modeling and simulation approach to estimating population increase in any country or region. When it comes to evolutionary algorithms, Montiel et al. (2009) presented a clever strategy for regulating population size. Using a method influenced by

[a]anuragpaliwal1993@yahoo.com, [b]ssyadavncs@gmail.com

fuzzy set theory, Sasu (2010) predicted Romania's population size. An example of a fuzzy differential equation defining a prey-predator paradigm was studied by Pandit and Singh (2014). With the use of Montroll's method, Cabrera et al. (2016) created a case study comparing data on the Peruvian population from 1961 to 2003. (supplied by INEI). When it comes to solving first-order fuzzy logistic equations, Zulkefli et al. (2017) developed an extended Runge-Kutta fourth-order method that uses estimated parameters to arrive at a numerical solution. Using the hypothesis that the model "propagates the fuzziness of the independent variable to the dependent variable," Nurkholipah et al. (2017) considered a fuzzy beginning value for the crisp model. When examining the Allee effect within the framework of Verhulst's logistic population model, Amarti et al. (2018). The Allee effect is a phenomenon in biology that shows there is a robust correlation between population size or density and the mean fitness of the population as a whole. A fuzzy initial value problem (FIVP) was investigated by Pedro et al. (2019) to characterize an auto correlated evolutionary process. The dynamics, such as boundedness, global asymptotic stability, and persistence of a positive fuzzy solution, were investigated by Zhang et al. (2022). They provided two case studies to illustrate the usefulness of the findings. It was recommended by Bressan and Stiegelmeier (2023) to examine the effectiveness of intervention techniques including lockdown, partial lockdown, and no-lockdown in containing the severe epidemic of COVID-19 in Brazil.

2. Steps of fuzzy-based model for human population growth

2.1 Definition of input and output variables

Input variable								Output variable	
Income (I)		Education (E)		Participation (P)		Age (A)		Growth rate (GR)	
IL	Low	EL	Low	PL	Low	AL	Low	GRVL	Very low
IM	Medium	EM	Medium	PM	Medium	AM	Medium	GRL	Low
IH	High	EH	High	PH	High	AH	High	GRM	Medium
								GRH	High
									Very high

2.2 Defining the input and output membership functions

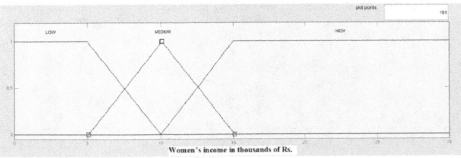

Figure 26.1 Input membership function of women's income

$$\mu_{IL} = \begin{cases} 1 & 0 \le x \le 5 \\ -0.2x + 2 & 5 \le x \le 10 \end{cases}, \mu_{IM} = \begin{cases} 0.2x - 1 & 5 \le x \le 10 \\ -0.2x + 3 & 10 \le x \le 15 \end{cases}, \mu_{IH}$$
$$= \begin{cases} 0.2x - 2 & 10 \le x \le 15 \\ 1 & 15 \le x \le 30 \end{cases}$$

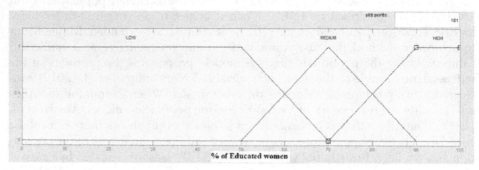

% of Educated women

Figure 26.2 Input membership function of % of educated women

$$\mu_{EL} = \begin{cases} 1 & 0 \le x \le 50 \\ -0.05x + 3.5 & 50 \le x \le 70 \end{cases}, \mu_{EM} = \begin{cases} 0.05x - 2.5 & 50 \le x \le 70 \\ -0.05x + 4.5 & 70 \le x \le 90 \end{cases}, \mu_{EH}$$
$$= \begin{cases} 0.05x - 3.5 & 70 \le x \le 90 \\ 1 & 90 \le x \le 100 \end{cases}$$

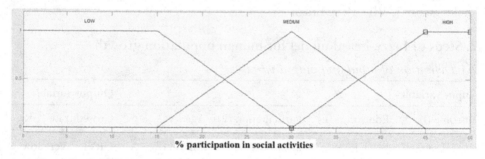

% participation in social activities

Figure 26.3 Input membership function of % participation in social activities

$$\mu_{PL} = \begin{cases} 1 & 0 \le x \le 15 \\ -0.6x + 2 & 15 \le x \le 30 \end{cases}, \mu_{PM} = \begin{cases} 0.6x - 1 & 15 \le x \le 30 \\ -0.6x + 3 & 30 \le x \le 45 \end{cases}, \mu_{PH}$$
$$= \begin{cases} 0.6x - 2 & 30 \le x \le 45 \\ 1 & 45 \le x \le 50 \end{cases}$$

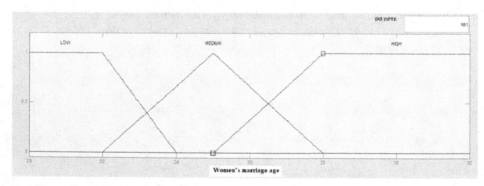

Women's marriage age

Figure 26.4 Input membership function of women's marriage age

$$\mu_{AL} = \begin{Bmatrix} 1 & 20 \leq x \leq 22 \\ -0.5x + 12 & 22 \leq x \leq 24 \end{Bmatrix}, \mu_{AM} = \begin{Bmatrix} 0.3x - 7.33 & 22 \leq x \leq 25 \\ -0.3x + 9.33 & 25 \leq x \leq 28 \end{Bmatrix}, \mu_{AH}$$
$$= \begin{Bmatrix} 0.3x - 8.33 & 25 \leq x \leq 28 \\ 1 & 28 \leq x \leq 32 \end{Bmatrix}$$

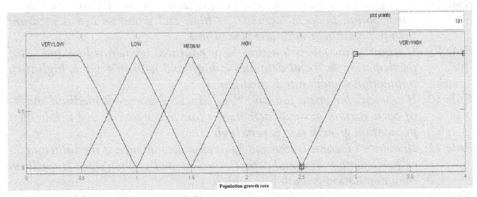

Figure 26.5 Output membership function of population growth rate

$$\mu_{GRVL} = \begin{Bmatrix} 1 & 0 \leq x \leq 0.5 \\ -2x + 2 & 0.5 \leq x \leq 1 \end{Bmatrix}, \mu_{GRL} = \begin{Bmatrix} 2x - 1 & 0.5 \leq x \leq 1 \\ -2x + 3 & 1 \leq x \leq 1.5 \end{Bmatrix}, \mu_{GRM}$$
$$= \begin{Bmatrix} 2x - 2 & 1 \leq x \leq 1.5 \\ -2x + 4 & 1.5 \leq x \leq 2 \end{Bmatrix}$$

$$\mu_{GRH} = \begin{Bmatrix} 2x - 3 & 1.5 \leq x \leq 2 \\ -2x + 5 & 2 \leq x \leq 2.5 \end{Bmatrix}, \mu_{GRVH} = \begin{Bmatrix} 2x - 5 & 2.5 \leq x \leq 3 \\ 1 & 3 \leq x \leq 4 \end{Bmatrix}$$

2.3 An explanation of fuzzy rules
The fuzzy population model's rules of inference are as follows:

Rule 1: If women's income is **low** and % of educated women is **low** and % of participation in social activities is **low** and women's age is **low** then population growth rate is **very high**

Rule 2: If women's income is **low** and % of educated women is **low** and % of participation in social activities is **low** and women's age is **medium** then population growth rate is **very high**

Rule 3: If women's income is **low** and % of educated women is **low** and % of participation in social activities is **low** and women's age is **high** then population growth rate is **high**

Rule 4: If women's income is **low** and % of educated women is **low** and % of participation in social activities is **medium** and women's age is **low** then population growth rate is **very high**

Rule 5: If women's income is **low** and % of educated women is **low** and % of participation in social activities is **medium** and women's age is **medium** then population growth rate is **high**

Rule 6: If women's income is **low** and % of educated women is **low** and % of participation in social activities is **medium** and women's age is **high** then population growth rate is **medium**

Rule 7:   If women's income is **low** and % of educated women is **low** and % of participation in social activities is **high** and women's age is **low** then population growth rate is **high**

Rule 8:   If women's income is **low** and % of educated women is **low** and % of participation in social activities is **high** and women's age is **medium** then population growth rate is **medium**

Rule 9:   If women's income is **low** and % of educated women is **low** and % of participation in social activities is **high** and women's age is **high** then population growth rate is **medium**

Rule 10:   If women's income is **low** and % of educated women is **medium** and % of participation in social activities is **low** and women's age is **low** then population growth rate is **very high**

Rule 11:   If women's income is **low** and % of educated women is **medium** and % of participation in social activities is **low** and women's age is **medium** then population growth rate is **high**

Rule 12:   If women's income is **low** and % of educated women is **medium** and % of participation in social activities is **low** and women's age is **high** then population growth rate is **medium**

Rule 13:   If women's income is **low** and % of educated women is **medium** and % of participation in social activities is **medium** and women's age is **low** then population growth rate is **high**

Rule 14:   If women's income is **low** and % of educated women is **medium** and % of participation in social activities is **medium** and women's age is **medium** then population growth rate is **medium**

Rule 15:   If women's income is **low** and % of educated women is **medium** and % of participation in social activities is **medium** and women's age is **high** then population growth rate is **medium**

Rule 16:   If women's income is **low** and % of educated women is **medium** and % of participation in social activities is **high** and women's age is **low** then population growth rate is **medium**

Rule 17:   If women's income is **low** and % of educated women is **medium** and % of participation in social activities is **high** and women's age is **medium** then population growth rate is **medium**

Rule 18:   If women's income is **low** and % of educated women is **medium** and % of participation in social activities is **high** and women's age is **high** then population growth rate is **medium**

Rule 19:   If women's income is **low** and % of educated women is **high** and % of participation in social activities is **low** and women's age is **low** then population growth rate is **high**

Rule 20:   If women's income is **low** and % of educated women is **high** and % of participation in social activities is **low** and women's age is **medium** then population growth rate is **medium**

Rule 21:   If women's income is **low** and % of educated women is **high** and % of participation in social activities is **low** and women's age is **high** then population growth rate is **medium**

Rule 22: If women's income is **low** and % of educated women is high and % of participation in social activities is medium and women's age is low then population growth rate is medium

Rule 23: If women's income is **low** and % of educated women is **high** and % of participation in social activities is **medium** and women's age is **medium** then population growth rate is **medium**

Rule 24: If women's income is **low** and % of educated women is **high** and % of participation in social activities is **medium** and women's age is **high** then population growth rate is **medium**

Rule 25: If women's income is **low** and % of educated women is **high** and % of participation in social activities is **high** and women's age is **low** then population growth rate is **medium**

Rule 26: If women's income is **low** and % of educated women is **high** and % of participation in social activities is **high** and women's age is **medium** then population growth rate is **medium**

Rule 27: If women's income is **low** and % of educated women is **high** and % of participation in social activities is **high** and women's age is **high** then population growth rate is **low**

Rule28: If women's income is **medium** and % of educated women is **low** and % of participation in social activities is **low** and women's age is **low** then population growth rate is **very high**

Rule 29: If women's income is **medium** and % of educated women is **low** and % of participation in social activities is **low** and women's age is **medium** then population growth rate is **high**

Rule 30: If women's income is medium and % of educated women is low and % of participation in social activities is low and women's age is hi then population growth rate is medium

Rule 31: If women's income is **medium** and % of educated women is **low** and % of participation in social activities is **medium** and women's age is **low** then population growth rate is **high**

Rule 32: If women's income is **medium** and % of educated women is **low** and % of participation in social activities is **medium** and women's age is **medium** then population growth rate is **medium**

Rule 33: If women's income is **medium** and % of educated women is **low** and % of participation in social activities is **medium** and women's age is hi then population growth rate is **medium**

Rule 34: If women's income is **medium** and % of educated women is **low** and % of participation in social activities is **high** and women's age is **low** then population growth rate is **medium**

Rule 35: If women's income is **medium** and % of educated women is **low** and % of participation in social activities is **high** and women's age is **medium** then population growth rate is **medium**

Rule 36: If women's income is **medium** and % of educated women is **low** and % of participation in social activities is **high** and women's age is **high** then population growth rate is **medium**

Rule 37:   If women's income is **medium** and % of educated women is **medium** and % of participation in social activities is **low** and women's age is **low** then population growth rate is **high**

Rule 38:   If women's income is **medium** and % of educated women is **medium** and % of participation in social activities is **low** and women's age is **medium** then population growth rate is **medium**

Rule 39:   If women's income is **medium** and % of educated women is **medium** and % of participation in social activities is **low** and women's age is **high** then population growth rate is **medium**

Rule40:   If women's income is **medium** and % of educated women is **medium** and % of participation in social activities is **medium** and women's age is **low** then population growth rate is **medium**

Rule 41:   If women's income is **medium** and % of educated women is **medium** and % of participation in social activities is **medium** and women's age is **medium** then population growth rate is **medium**

Rule 42:   If women's income is **medium** and % of educated women is **medium** and % of participation in social activities is **medium** and women's age is **high** then population growth rate is **medium**

Rule 43:   If women's income is **medium** and % of educated women is **medium** and % of participation in social activities is **high** and women's age is **low** then population growth rate is **medium**

Rule 44:   If women's income is **medium** and % of educated women is **medium** and % of participation in social activities is **high** and women's age is **medium** then population growth rate is **medium**

Rule 45:   If women's income is **medium** and % of educated women is **medium** and % of participation in social activities is **high** and women's age is **high** then population growth rate is **low**

Rule 46:   If women's income is **medium** and % of educated women is **high** and % of participation in social activities is **low** and women's age is **low** then population growth rate is **medium**

Rule 47:   If women's income is **medium** and % of educated women is **high** and % of participation in social activities is **low** and women's age is **medium** then population growth rate is **medium**

Rule 48:   If women's income is **medium** and % of educated women is **high** and % of participation in social activities is **low** and women's age is **high** then population growth rate is **medium**

Rule 49:   If women's income is **medium** and % of educated women is **high** and % of participation in social activities is **medium** and women's age is **low** then population growth rate is **medium**

Rule 50:   If women's income is **medium** and % of educated women is **high** and % of participation in social activities is **medium** and women's age is **medium** then population growth rate is **medium**

Rule 51:   If women's income is **medium** and % of educated women is **high** and % of participation in social activities is **medium** and women's age is **high** then population growth rate is **low**

Rule 52: *If women's income is **medium** and % of educated women is **high** and % of participation in social activities is **high** and women's age is **low** then population growth rate is **medium***

Rule 53: *If women's income is **medium** and % of educated women is **high** and % of participation in social activities is **high** and women's age is **medium** then population growth rate is **low***

Rule 54: *If women's income is **medium** and % of educated women is **high** and % of participation in social activities is **high** and women's age is **high** then population growth rate is **very low***

Rule 55: *If women's income is **high** and % of educated women is **low** and % of participation in social activities is **low** and women's age is **low** then population growth rate is **high***

Rule 56: *If women's income is **high** and % of educated women is **low** and % of participation in social activities is **low** and women's age is **medium** then population growth rate is **medium***

Rule 57: *If women's income is **high** and % of educated women is **low** and % of participation in social activities is **low** and women's age is **high** then population growth rate is **medium***

Rule 58: *If women's income is **high** and % of educated women is **low** and % of participation in social activities is **medium** and women's age is **low** then population growth rate is **medium***

Rule 59: *If women's income is **high** and % of educated women is **low** and % of participation in social activities is **medium** and women's age is **medium** then population growth rate is **medium***

Rule 60: *If women's income is **high** and % of educated women is **low** and % of participation in social activities is **medium** and women's age is **high** then population growth rate is **medium***

Rule 61: *If women's income is **high** and % of educated women is **low** and % of participation in social activities is **high** and women's age is **low** then population growth rate is **medium***

Rule 62: *If women's income is **high** and % of educated women is **low** and % of participation in social activities is **high** and women's age is **medium** then population growth rate is **medium***

Rule 63: *If women's income is **high** and % of educated women is **low** and % of participation in social activities is **high** and women's age is **high** then population growth rate is **low***

Rule 64: *If women's income is **high** and % of educated women is **medium** and % of participation in social activities is **low** and women's age is **low** then population growth rate is **medium***

Rule 65: *If women's income is **high** and % of educated women is **medium** and % of participation in social activities is **low** and women's age is **medium** then population growth rate is **medium***

Rule 66: *If women's income is **high** and % of educated women is **medium** and % of participation in social activities is **low** and women's age is **high** then population growth rate is **medium***

Rule 67: If women's income is **high** and % of educated women is **medium** and % of participation in social activities is **medium** and women's age is **low** then population growth rate is **medium**

Rule 68: If women's income is **high** and % of educated women is **medium** and % of participation in social activities is **medium** and women's age is **medium** then population growth rate is **medium**

Rule 69: If women's income is **high** and % of educated women is **medium** and % of participation in social activities is **medium** and women's age is **high** then population growth rate is **low**

Rule 70: If women's income is **high** and % of educated women is **medium** and % of participation in social activities is **high** and women's age is **low** then population growth rate is **medium**

Rule 71: If women's income is **high** and % of educated women is **medium** and % of participation in social activities is **high** and women's age is **medium** then population growth rate is **low**

Rule 72: If women's income is **high** and % of educated women is **medium** and % of participation in social activities is **high** and women's age is **high** then population growth rate is **very low**

Rule 73: If women's income is **high** and % of educated women is **high** and % of participation in social activities is **low** and women's age is **low** then population growth rate is **medium**

Rule 74: If women's income is **high** and % of educated women is **high** and % of participation in social activities is **low** and women's age is **medium** then population growth rate is **medium**

Rule 75: If women's income is **high** and % of educated women is **high** and % of participation in social activities is **low** and women's age is **high** then population growth rate is **low**

Rule 76: If women's income is **high** and % of educated women is **high** and % of participation in social activities is **medium** and women's age is **low** then population growth rate is **medium**

Rule 77: If women's income is **high** and % of educated women is **high** and % of participation in social activities is **medium** and women's age is **medium** then population growth rate is **low**

Rule 78: If women's income is **high** and % of educated women is **high** and % of participation in social activities is **medium** and women's age is high then population growth rate is **very low**

Rule 79: If women's income is **high** and % of educated women is **high** and % of participation in social activities is **high** and women's age is **low** then population growth rate is **low**

Rule 80: If women's income is **high** and % of educated women is **high** and % of participation in social activities is **high** and women's age is **medium** then population growth rate is **very low**

Rule 81: If women's income is **high** and % of educated women is **high** and % of participation in social activities is **high** and women's age is **high** then population growth rate is **very low**

3. Results and dsicussion

3.1 *Estimating future population size*

Population growth rate for different input variables.

S.No.	Women's Income in thousands of Rs.	% of educated women	% participation in social activities	Women's marriage age.	Population growth rate
1	15	50	25	26	1.51
2	20	70	50	30	0.373
3	30	40	50	30	1
4	30	40	50	25	1.5
5	6	60	50	20	1.75
6	15	20	20	20	1.88
7	15	40	50	28	1.01
8	20	90	20	28	0.777
9	25	70	20	30	1.32
10	6	10	10	20	3.35
11	15	10	10	20	2.06
12	15	40	40	29	1.18
13	12	70	50	29	0.74
14	12	70	20	30	1.31
15	8	20	20	30	1.71
16	8	75	50	30	1.03
17	8	75	50	24	1.34
18	8	10	10	24	2.76
19	7	35	35	23	2.69
20	7	35	50	23	1.78

3.2 *3D Surface plots*

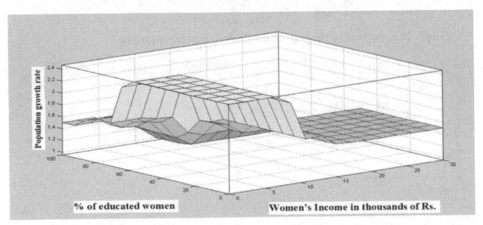

Graph 26.1 Surface of the population growth rate fuzzy output

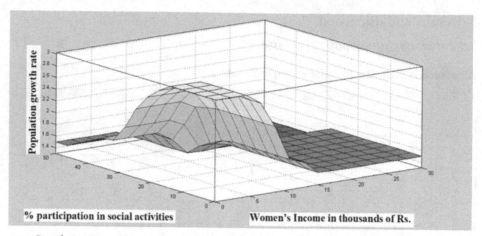

Graph 26.2 Surface of the population growth rate fuzzy output

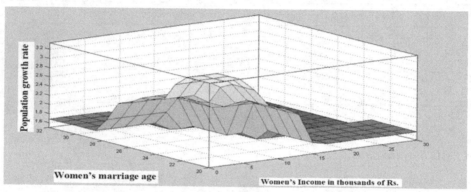

Graph 26.3 Surface of the population growth rate fuzzy output

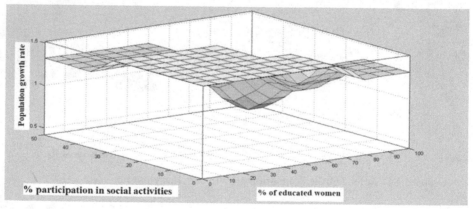

Graph 26.4 Surface of population growth rate fuzzy output

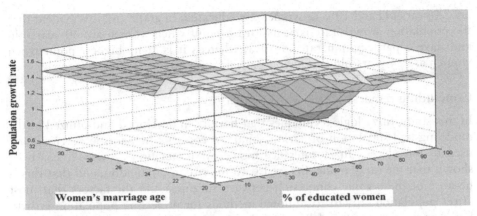

Graph 26.5 Surface of population growth rate fuzzy output

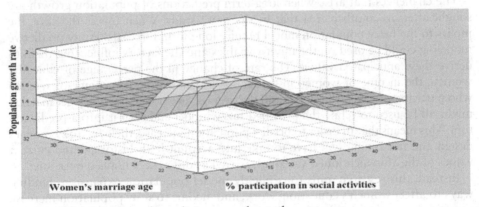

Graph 26.6 Surface of population growth rate fuzzy output

Graphs 26.1–26.6 shows the surface plot of a standard fuzzy inference system's population growth rate for a variety of input variables. Between input and output variables, 3D surfaces have been drawn. The response surface is a technique used to show how the given variables behave on a three-dimensional surface. As shown in graph (1), the population growth rate is at its lowest when the percentage of educated women is 70 and the average female income is 20,000, while it is at its highest when the percentage of educated women and the average female income are, respectively, 10 and 8,000. According to graph (2), the population growth rate is at its lowest when women participate in social activities at a rate of 50% and earn 20,000, while it is at its highest when women participate at a rate of 10% and earn 8,000, respectively. We can observe from graph (3) that the population growth rate is at its lowest point when women are 30-years-old and earn 20,000, and at its highest point when women are 24-years-old and earn 8,000, respectively. As shown in graph (4), population growth rate is least when 50% of women participate in social activities and 70% of women have college degrees, while growth rate is largest when these percentages are equal to 10% and 10%, respectively. The population growth rate is at its lowest in graph (5) when women are 30-years-old and 70% of women are educated and at its highest when women are 24-years-old and 10% of women are educated. In graph (6), it is evident that the population growth rate is at its lowest point when women are 30-years-old and have a participation rate of 50%, whereas it is at its highest point when women are 24-years-old and have a participation rate of 10%, respectively.

Concluding remarks

In this study, fuzzy logic techniques that have been developed and utilized in the modeling and simulation of various kinds of complex humanistic processes are used to introduce a fuzzy inference model for population growth. It has been shown that the fuzzy model fits well with the rise in population, and that this is because of the organic connection between fuzziness and the factors that most affect population growth. Traditional methods for projecting population growth based on demographic models and stochastic approaches are notoriously complex.

The challenges that arise when long-term predictions of population growth are sought after are another area that merits more study. When this is the case, the inputs to the fuzzy population model require long-term forecasts. Fuzzy predictive modeling could be used on the input variables and the factors influencing them to derive these anticipated values. The rate of change in the percentage of educated women, the income of women in thousands of rupees, the percentage of women who participate in social activities and the average age at which women get married might all be used to refine the provided fuzzy model and turn it into a prediction tool. This means that in order to predict population increase, the new fuzzy model's design and fine-tuning will have to take into account both historical and current values of the aforementioned variables, as well as differences noticed over time.

Instead, this research aims to explicitly model population expansion. Finding ways to affect women's education, work, and income to slow population increase may be a secondary goal. After all, this study can be viewed as a good attempt aiming at setting the foundation of and inspiring more research in the domain of fuzzy modeling as it relates to demographic issues.

References

1. Amarti, Z., Nurkholipah, N. S., Anggriani, N., and Supriatna, A. K. (2018). Numerical solution of a logistic growth model for a population with Allee effect considering fuzzy initial values and fuzzy parameters. *Mat. Sci. Engg.*, 332, 012051.
2. Barros, L. C., Bassanezi, R. C., and Tonelli, P. A. (2000). Fuzzy modelling in population dynamics. *Ecol. Model.*, 128, 27–33.
3. Bressan, G. M. and Stiegelmeier, E. W. (2023). Fuzzy modelling on the evolution of COVID-19 epidemic under the effects of intervention measures. *Brazilian Arch. Biol. Technol.*, 66, e23220425.
4. Cabrera, N. V., Jafelice, R. S. M., and Bertone, A. M. A. (2016). Montroll's model applied to a population growth data set using type-1 and type-2 fuzzy parameters. *Biomatematica*, 26, 145–160.
5. Diab, H. and Sadde, J. (2005). Fuzzy inference modeling methodology for the simulation of population growth. *Int. Arab J. Inform. Technol.*, 2(1), 75–86.
6. Montiel1, O., Castillo, O., Melin, P., and Sepulveda, R. (2009). Mediative fuzzy logic for controlling population size in evolutionary algorithms. *Intel. Inform. Manag.*, 1, 108–119.

7.   Nurkholipah, N. S., Amarti, Z., and Supriatna, A. K. (2017). A fuzzy mathematical model of West Java population with logistic growth model. Mat. Sci. Engg., 332, 012035.

8.   Pandit, P. and Singh, P. (2014). Prey predator model with fuzzy initial conditions. *Int. J. Engg. Innov. Technol. (IJEIT)*, 3(12), 65–68.

9.   Pedro, F. S., Barros de, L. C., and Esmi, E. (2019). Population growth model via interactive fuzzy differential equation. *Inform. Sci.*, 481, 160–173.

10. Sasu, A. (2010). An application of fuzzy time series to the Romanian population. *Series III: Math. Inform. Phy.*, 3(52), 125–132.

11. Zhang, Q., Ouyang, M., and Zhang, Z. (2022). On second-order fuzzy discrete population model. *Open Math.*, 20, 125–139.

12. Zulkefli, N. A. I., Hoe, Y. S., and Mann, N. (2017). The application of fuzzy logistic equations in population growth with parameter estimation via minimization. *Malaysian J Fundam. Appl. Sci.*, 13(2), 109–112.

Printed in the United States
by Baker & Taylor Publisher Services